# MARS

**From 4.5 billion years ago to the present**

**Dedication**

*Michael Hanlon (1964–2016) had a particular fondness for the planet Mars*

# Acknowledgements

I would like to thank, in no particular order, Steve Rendle, W. David Woods, James Joel Knapper, Wes Huntress, Giancarlo Genta, Tom Ruen, Michael Diggles, Dutch von Ehrenfried, David Baker, Marc Rayman, and Bill Sheehan.

# Nomenclature

The nomenclature of the International Astronomical Union includes the following features on Mars:

| Chaos | A distinctive area of broken or jumbled terrain. |
|---|---|
| Chasma, chasmata | A deep, elongated, steep-sided depression. |
| Crater, craters | A circular depression. |
| Labyrinthus, labyrinthi | A complex of intersecting valleys or ridges. |
| Lacus, lacūs | A small plain. |
| Mare, maria | A dark albedo area. |
| Mons, montes | A mountain or (plural) a mountain range. |
| Oceanus | Very large low-lying area. |
| Patera, paterae | An irregular crater, or a complex one with scalloped edges. It usually refers to the dish-shaped depression atop a volcano. |
| Planitia, planitiae | A low-lying plain. |
| Planum, plana | A plateau or high plain. |
| Sinus, sinūs | A small plain. |
| Terra, terrae | An extensive land mass. |
| Tholus, tholi | A small domical mountain or hill. |
| Vallis, valles | A valley. |
| Vastitas, vastitates | An extensive plain. |

# MARS

## From 4.5 billion years ago to the present

# Owners' Workshop Manual

An insight into the study and exploration of
the Red Planet

**David M. Harland**

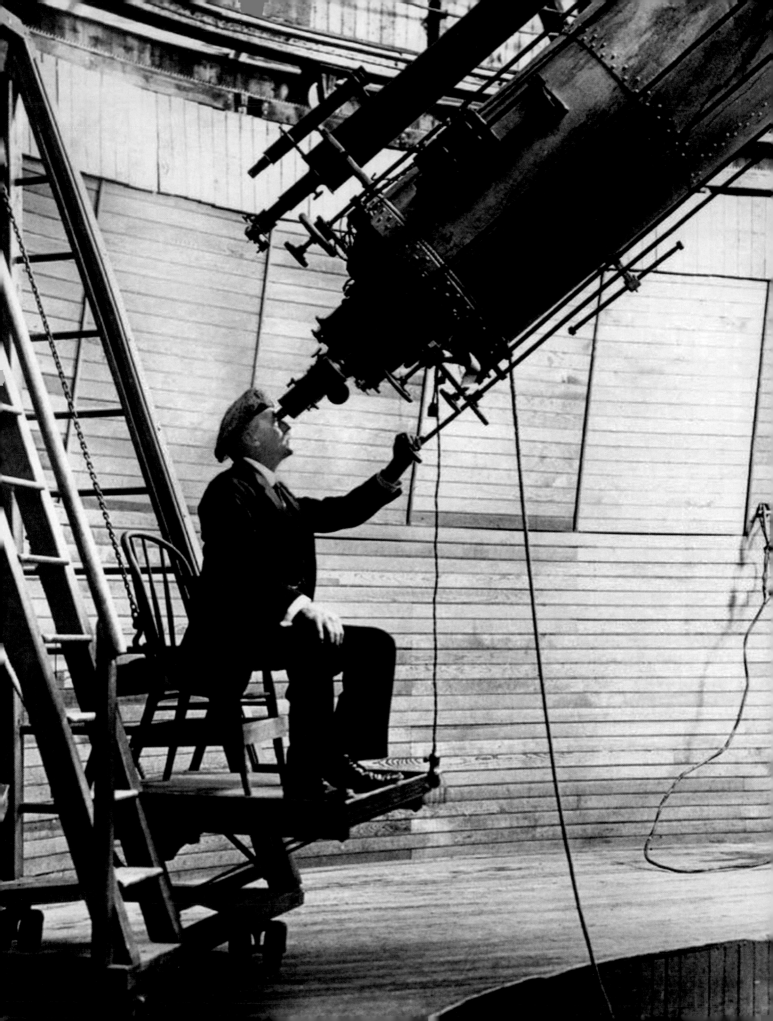

# Contents

**OPPOSITE** In the final years of the 19th century, Percival Lowell equipped a private observatory with a powerful telescope specifically to study the planet Mars. *(Lowell Observatory)*

# Introduction

This manual reviews what we have discovered about Mars since Johannes Kepler in the first decade of the 17th century. Over that time, our impression of the planet has changed from it being only a light that travels across the sky to a fascinating landscape that our robotic proxies are investigating while we create the means to go there ourselves.

When astronomers observed Mars they often took it for granted that the planet was inhabited. Indeed, in his scientific papers, William Herschel referred almost casually to the natives. When Percival Lowell spectacularly announced that an ancient race had built canals to transport water from the polar ice caps into the arid deserts, this proposition did not seem outlandish to members of the public.

When H. G. Wells wrote a dramatic tale of Martians invading Earth, lured by its oceans, there seemed little prospect that one day we might send vehicles to Mars.

The Soviet Union was first to attempt this feat but had little success. The Americans fared better, and later other nations joined in. But it isn't an easy task, and a fair proportion of missions have failed. The story includes the tremendous excitement of viewing the first picture transmitted from the surface of the planet and the frustration of navigating a probe through space only to lose it just a few hours or minutes short of its target.

Over time, our expectations of a given mission have increased and several have proven to be outstandingly successful. Most notably, the Opportunity rover which landed in early 2004 was still operating when this book was being written 13 years later.

As the virtual reality tools used on Earth to control rovers on Mars become ever more sophisticated it should become possible for members of the public to log into a downlink and observe the operations of a rover in real time. When an astronaut first sets foot on Mars, millions of people at home are sure to share in that momentous event as a total immersion experience.

It has yet to be determined whether there is microbial life on Mars but enticing evidence exists and perhaps soon we will know for sure. If life can be shown to have developed independently on at least two bodies in the solar system, that will suggest that life, as a chemical process, will develop wherever the conditions are conducive and it is likely to occur throughout the universe. Hence the results of our investigations may well be profound.

At the dawn of the Space Age, a slim volume could describe everything that was known about Mars. Since then the pace of discovery has been so rapid that this book can merely survey the vast scope of developments. In seeking to show the great variety of data that modern orbiters and surface missions can provide, I have maintained a hectic pace.

I start by summarising investigations through to 1877, explaining how the observations of that year influenced our thinking about Mars for half a century.

The early years of the Space Age are covered next, when our impressions were revised several times. Particular attention is given to the first landings and the experiments that sought evidence of life. Then the orbital investigations and parallel surface operations are brought up to date.

After briefly highlighting the role of Mars in science fiction, I end by looking ahead to the prospects for human missions. Instead of distributing information about missions throughout the text, this is presented in a chronology at the back of the book, along with some facts and figures about the planet and suggestions for further reading.

So let's get started…

**OPPOSITE Mars, which has a land surface equivalent to that of Earth, is being explored by robots; in this example, the Curiosity rover.** (NASA/ JPL-Caltech/MSSS)

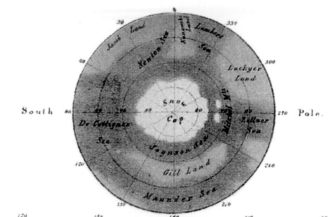

# CHART OF

*From drawings at Madeira in 1877.*

South 60 Pole.

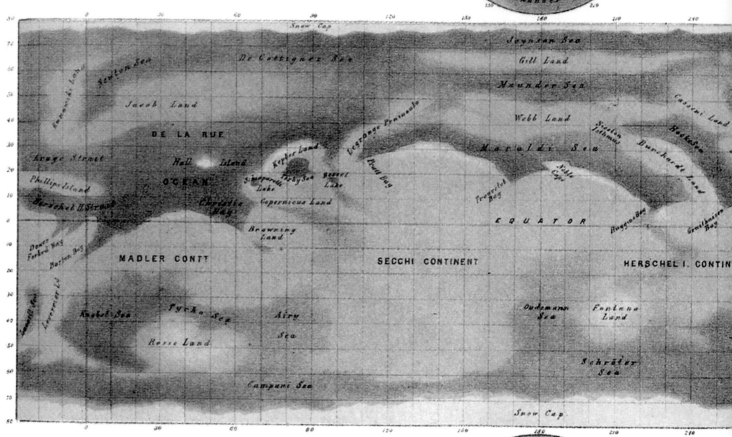

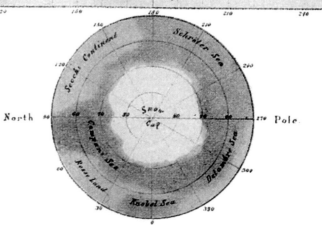

North Pole.

The details of this chart have been compared with views of the planet
by Schiaparelli, Trouvelot, Terby, De la Rue, Lockyer, Knobel, Christie,
Maunder, Brett Prover, and others. No form is introduced that has not
been confirmed by the drawings of at least three observers

MARS
by N.E.Green.

*The northern portion of the chart is supplied from drawings made in 1873 and may be considered provisional. The names with a few exceptions are those employed by Mr Proctor in his chart of Mars, adapted as far as possible to the recent observations. New names have been added by the author where required.*

## Chapter One

# Early studies of Mars

This chapter starts with Johannes Kepler in the early 17th century, who analysed the motions of Mars in the sky. The invention of the telescope resolved the planet as a disk, revealing the presence of polar ice caps, ochre tracts, and a number of dark patches that enabled its rotation rate to be measured and simple maps to be drawn.

**OPPOSITE** The map of Mars by Nathaniel Everett Green after observing the planet during the very favourable opposition of 1877. *(RAS)*

## BLOOD-RED MARS

As we now know, the reddish hue of Mars results from iron in the soil reacting with oxygen to produce rust. A similar reaction occurs in the haemoglobin that makes blood red. So Mars really is blood-red.

## ELLIPTICAL ECCENTRICITY

Eccentricity is the ratio of the distance between the centre of an ellipse and each focal point to the length of the semi-major axis.

The eccentricity of an ellipse therefore lies between zero (a circle) and one (a parabola).

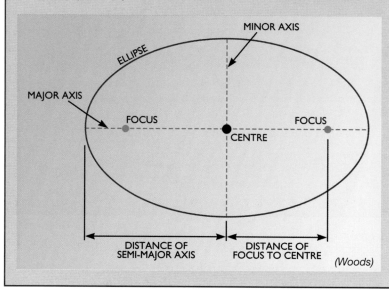

(Woods)

BELOW The 'equal areas in equal time' law of planetary motion devised by Johannes Kepler. (Woods)

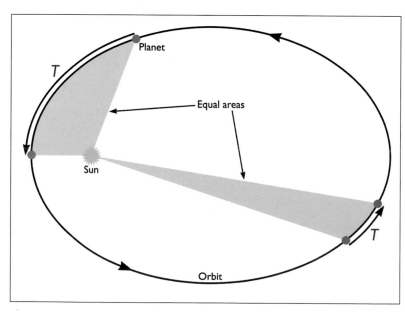

## Celestial motions

**W**hen Mars is high in the sky in the darkest hours of the night, it blazes down like a blood-red demon. Hence it is hardly surprising that in mythological lore, it is an ill omen. To the Babylonians it was Nirgal, the Star of Death. To the ancient Greeks it was Ares, the God of War, whom the Romans later called Mars. Nowadays it is often casually referred to simply as the Red Planet.

As a point-like object that moves against the background of 'fixed' stars, it is a 'planet' (which means wanderer).

In 1543, the year of his death, the Polish mathematician Nicolaus Copernicus published an account of how he believed the planets, including Earth, travelled in circles around the Sun, thereby overturning the old notion that everything circled Earth.

With the support of King Frederick II of Denmark, Tycho Brahe established an 'observatory' on an island in the channel off Copenhagen to map the positions of the stars and the motions of the planets with unprecedented accuracy. In 1599, Brahe relocated to Prague to continue under the patronage of Emperor Rudolph II.

Upon Brahe's death in 1601, his unique archive of observations passed to Johannes Kepler who, in recent years, had been his principal assistant. At that time, the distinction between astronomy and astrology was so fuzzy that when Kepler set out to use Brahe's data to undertake a mathematical analysis of the motion of Mars, he subsidised his income by casting horoscopes.

In 1609 Kepler announced that the orbit of Mars is an ellipse possessing an eccentricity of 0.0935.

As Kepler pursued his analysis further, he identified three laws of planetary motion.

The first law of planetary motion states that the orbit of a planet is an ellipse that has the Sun at one of its focal points; the second focal point is vacant.

The speed of a body in a circular orbit would be uniform, but it would vary for an ellipse. The second law says the rate at which a planet travels is fastest when nearest the Sun and slowest when farthest away. In the graphical scheme which Kepler employed, he was able to determine the speed of a planet at any position because the area in the ellipse that is swept by the line which links it to the Sun in any interval of time is proportional to that interval. This is commonly described as the law of 'equal areas in equal times'.

The third law says that in relative terms, the squares of the times of revolution of two planets are proportional to the cubes of their mean distances from the Sun.

These laws are empirical because they state *how* planets orbit the Sun but do not explain *why*. In due course, Isaac Newton in England explained them in terms of the force of universal gravitation.

Kepler's laws apply to Earth in orbit around the Sun, satellites orbiting planets, and the Moon orbiting Earth.

As would become evident much later, Mars orbits the Sun about 50% farther out than does Earth; its heliocentric distance varies between 206 and 249 million km, with a mean of 228 million km. With an eccentricity of 0.0167, Earth remains within 2.5 million km of its mean heliocentric distance of 149.5 million km. As the planets pursue their orbits around the Sun in accordance with the empirical laws inferred by Kepler, Mars, being farther out, travels through space more slowly. In terms of

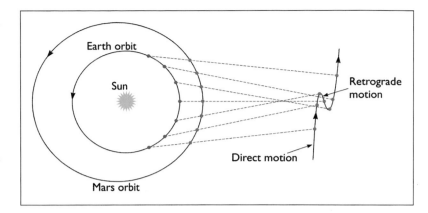

our calendar, Mars takes 687 days to make one revolution. The longer period and the fact that the plane in which Mars orbits the Sun is inclined 1.8° to that of Earth (called the 'ecliptic') causes it to trace out irregular loops on the sky, with its progressive eastward motion being disrupted by episodes of 'retrograde' motion.

Oppositions are instances when Mars is opposite the Sun in the sky, and they occur at intervals of 780 days. Successive oppositions occur at a different position along the Zodiac so that, after 8 oppositions of Mars, and with Earth having made 17 revolutions, Mars will have returned to the same part of its orbit. At a 'perihelic' opposition, the planet can close to within 56 million km of Earth; the range at an 'aphelic' opposition is much greater.

**ABOVE It was the pronounced 'looping' of Mars across the sky that enabled Johannes Kepler to devise his laws of planetary motion.** *(Woods)*

## MARS OPPOSITION AND CONJUNCTION

Earth and Mars pursue their independent orbits around the Sun. There are two cases where the Sun and planets can be in line: one when the planets lie on the same side of the Sun (opposition) and the other when they are on opposite sides of the Sun (conjunction).

When at opposition Mars is prominent in the night sky but at conjunction it is invisible in daylight.

**RIGHT How the oppositions, conjunctions and seasons of Mars are affected by its elliptical orbit.** *(Woods)*

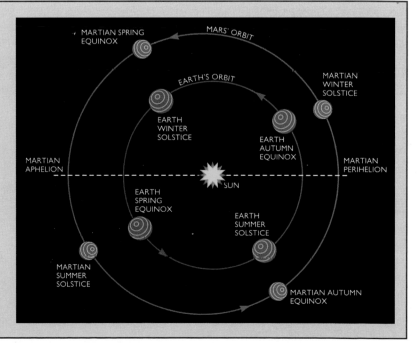

**BELOW A depiction of Christiaan Huygens using a very long focal length 'aerial' refractor. (INSET) Sketches of Mars by Christiaan Huygens in 1659 (top), 1672 and 1683.**

# Pioneering telescopic observers

In 1610, Galileo Galilei in Padua, Italy, turned an early telescope to Mars but was barely able to discern it as a disk.

Francesco Fontana, a lawyer in Naples, resolved the planet's disk in 1636, and two years later he noted that it was showing a gibbous

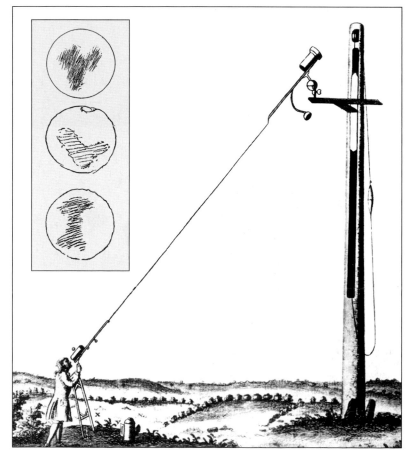

phase. Unfortunately, the individual features he perceived were optical flaws in his telescope.

The first person to reliably report markings on the disk was Neopolitan Jesuit Daniello Bartoli in 1644, but he did not draw them.

Giovanni Battista Riccioli, a Jesuit professor at the University of Bologna, and his student Francesco Grimaldi, saw 'albedo' (brightness) variations on the planet in 1651 and again at oppositions through to 1657, but they provided no drawings.

## A rotating globe

The first person to document what he saw on Mars was the Dutch lens-maker Christiaan Huygens. By timing when a prominent dark V-shaped feature crossed the centre of the disk between 28 November and 1 December 1659 he concluded "the days and nights are about the same length as ours". His measurement of the rotational period was the first discovery to be made about the nature of the planet itself, as opposed to its path across the terrestrial sky.

In Bologna in Italy, Giovanni Domenico Cassini observed Mars during the opposition of March 1666 and made a series of drawings showing tremendous variation in the features. He argued that if Earth were to be viewed from afar its oceans would appear dark and the continents bright, so it ought to be the same for Mars. Although this was an aphelic opposition and hence not well presented, Cassini was nevertheless able to refine the rotational period. By observing on a series of nights he noted that the planet rotated slightly slower than Earth. From the fact that it took 38 days for a given feature to appear at the same place at a given hour he calculated the period as 24hr 40min. So the local year comprises 669 sols (as the Martian day is nowadays termed).

Cassini reported a fuzzy white cap at the south pole in 1666, and Huygens saw it prominently at the perihelic opposition of 1672.

Young Giancomo Filippo Maraldi made a study of Mars in 1672, and again on subsequent favourable occasions. Regarding the general features he wrote, "The patches visible with large telescopes on the planet's disk are not normally well defined, and they often change in form, not only from one opposition to another … but even from one month to another." This implied that many of the albedo features were clouds in a stormy atmosphere. "Notwithstanding these changes, the patches last for long enough for us to follow them for a time sufficient for a determination of the rotational period." His results confirmed Cassini's period.

At the perihelic opposition of 1704, Maraldi observed that the southern cap underwent a slight revolution, meaning it was not centred precisely on the axial pole. He also observed that the areal extent of the cap varied over time. At the particularly favourable opposition of 1719, he noted that the southern cap was absent for a time. In 1716 he was the first to discern a small white patch at the north pole.

The variations of the polar caps suggested that Mars underwent seasonal variations.

This concluded the initial phase of the telescopic study of Mars because the perihelic oppositions in 1734, 1751 and 1766 all passed without comment. It was clear that Mars rotated slightly more slowly than Earth, had transient polar caps, and had some semi-permanent albedo features along with others that seemed to be so variable as to be atmospheric in nature.

# Early mapping

Whenever Mars was well presented between 1777 and 1783, Frederick William Herschel in England studied it using telescopes that he had constructed himself, resolving much more detail than had his predecessors.

Herschel's interests were the rotational period of the planet, the inclination of its rotational axis, and how the eccentric orbit and

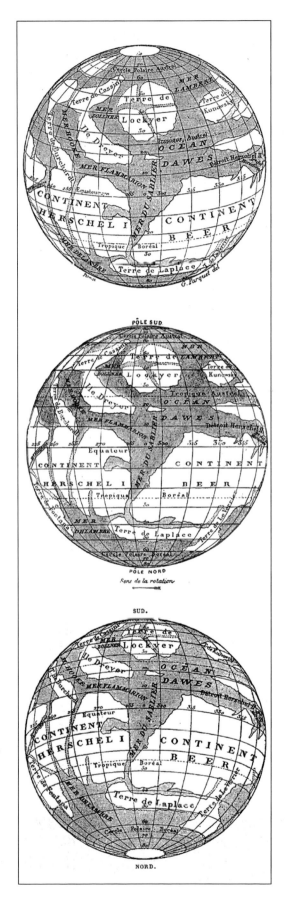

**LEFT** Views of one hemisphere of Mars by Camille Flammarion showing how the tilt of the planet's rotational axis relative to the line of sight from Earth presents a changing perspective, sometimes favouring one or other of the poles and sometimes neither.

the axial tilt combine to cause the variations of the polar caps. Although he made many sketches, he paid little attention to the more general albedo features.

Due to the inclination of the planet's rotational axis, sometimes one pole was presented for viewing, sometimes the opposite pole was presented, and at times it was possible to view both poles simultaneously.

At the perihelic opposition of 1781 Herschel verified Maraldi's finding that the southern cap was not centred at the axial pole and established that the northern cap was also misaligned. By analogy with Earth, he inferred from the way that the northern and southern hemispheres were alternately heated by the Sun (owing to the axial tilt) that the polar caps were fields of snow and ice which expanded in winter and retreated in summer.

By monitoring the passage of dark disk features from night to night, Herschel calculated the rotational period to be 24hr 39min 21.7sec.

On two occasions in October 1783 Herschel saw Mars pass in front of a star,

and inferred from the rapidity with which the starlight diminished that the planet's atmosphere was thin. However, on occasion he saw short-lived changes that he took to be atmospheric in nature.

Integrating what was then known about Mars, it was transformed from a point of light in the sky into a world in its own right. As the obliquity of its rotational axis was of a similar magnitude to that of Earth, the planet had a seasonal cycle, and because its orbit was elliptical its motion was fastest at perihelion. Consequently the four seasons were not of equal duration: the southern summer season lasted 156 sols but the southern winter lasted 177 sols. Since the planet received 44% more energy from the Sun at perihelion than at aphelion, the fact that the south pole was tilted sunward at perihelion implied that the southern summer was hot, whereas the southern winter, with the pole tilted away at aphelion, was harsh. In the north, the duration of the seasons was reversed and the variation in surface conditions was less dramatic. Presuming Mars to be inhabited, Herschel opined "the inhabitants probably enjoy conditions analogous to ours in several respects".

Johann Heinrich Mädler teamed up with Wilhelm Wolff Beer, a wealthy banker in Berlin with an interest in astronomy who had a private observatory, and they set out to study Mars at its perihelic opposition of 1830. Their goal was to produce a map of the planet. After several oppositions they had accumulated a large stack of drawings. Having deemed the atmosphere to be too thin to support a vigorous weather system, they inferred that the albedo features were surface detail. They took note of variations of the sizes, shapes and hues of the albedo features, and monitored the seasons. In publishing their map in 1840 they projected equatorial and polar views and assigned letters to identify features of interest. In effect Beer and Mädler introduced the subject of 'areography' as a counterpart to geography.

Having determined the rotational period to be 24hr 37min 23.7sec, Beer and Mädler resolved the 2min discrepancy with Herschel's period by re-examining his observations. They realised the planet had turned on its axis once more between 1777 and 1779 than he had thought.

**BELOW** Sketches of Mars by Wilhelm W. Beer and Johann H. Mädler at the start of their observations of the planet in 1830. *(With thanks to Bill Sheehan)*

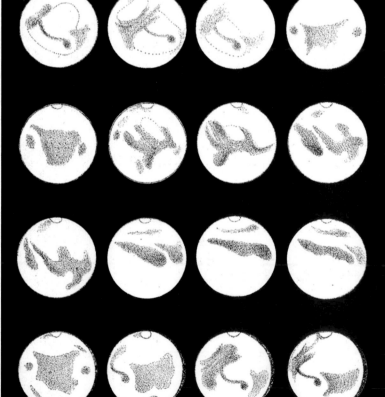

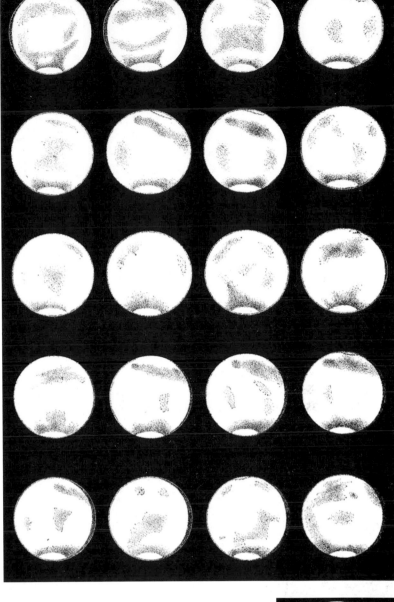

When this was corrected their times agreed to within a few seconds.

At the 1858 opposition, Pietro Angelo Secchi at the Collegio Romano noted a short-lived white feature. This was the first sighting of what would later be called 'white clouds'.

The opposition of 1860 was perihelic but too low in the sky for observers in Europe. The 1862 opposition was much better presented and Frederik Kaiser in Holland created a map that was a distinct improvement on the one by Beer and Mädler. By comparing his timings with those of Huygens in 1659 and Herschel in 1783, Kaiser refined the period to 24hr 37min 22.62sec. (The pioneering sketches made by Huygens really proved their worth!) In England, Joseph Norman Lockyer produced a set of exquisite drawings he later told the Royal Astronomical Society were in "marvellous agreement" with those of Beer and Mädler.

The dark features on Mars had generally been presumed to be seas, but in 1860 Emmanuel Liais, working in Paris, argued that large bodies of water were unlikely to undergo a seasonal cycle of darkening. Instead, he argued the dark areas

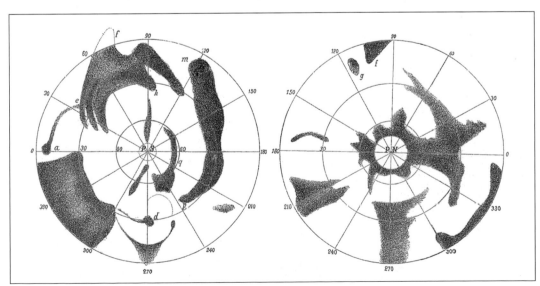

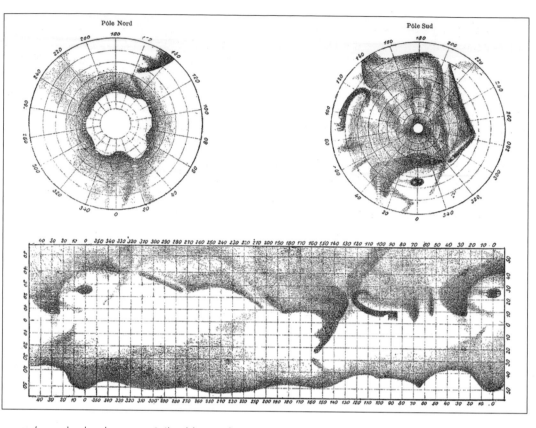

were *dry* seabeds where vegetation bloomed in reaction to the retreating polar caps infusing water vapour into the atmosphere. Secchi disagreed, saying the seasonal waxing and waning of the caps "can be explained only by a melting of the snow or a disappearance of the clouds" and as liquid water was "a natural result of the behaviour of the snows" it was evident to him that "the existence of seas and continents … has been conclusively proved". John Phillips, a geologist in England who had observed Mars extensively during the 1862 opposition, said that if the dark areas were open water then he should have seen the glint of the Sun, which he had not.

William Rutter Dawes created many drawings of Mars at the 1862 and 1864 oppositions. In 1867 Richard Anthony Proctor, a prolific author of popular books on astronomy, used these observations to compile a map that was considerably more detailed than that of Beer and Mädler. He went further by *naming* features after astronomers, living and dead, who had studied the planet. The prominent V-shaped feature first sketched by Huygens was named the Kaiser Sea. He used the small dark spot on the equator that Beer and Mädler had labelled 'A' for the meridian. However, Proctor attracted criticism for having given undue priority to English astronomers. Even worse he named an ocean, a sea, a strait, a bay, an island, and a continent in honour of his friend!

In 1870 Proctor noted that he could recognise features on drawings made by Robert Hooke in 1666, and used this 200-year time base to refine the rotational period to 24hr 37min 22.73sec. He also noted that, over that interval, an error of 0.1sec in the period would have resulted in a discrepancy of 2hr.

Camille Flammarion, in Paris, started observing Mars at its 1871 opposition. When he published a map in 1876 he retained the meridian feature, renaming it the Meridian Bay.

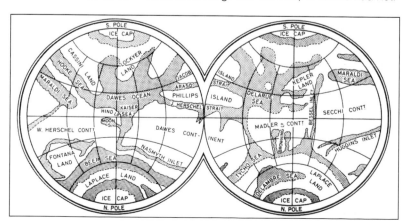

In *Mars and its Conditions of Habitability* published in 1892, Flammarion cited the nebular hypothesis as evidence that the planet had evolved more rapidly than Earth. On this basis he reasoned its surface was severely eroded and essentially flat, and the seasonal variations were inundations of the shorelines by water that was "of Mediterranean shallowness".

Following up in 1893 with *The Planet Mars*, Flammarion recalled a speculation from the previous century by Johann Heinrich Lambert, and suggested the ochre areas were infested by plants of a species that was unique to that planet. "Why is not the Martian vegetation green? Why should it be?" Pointing out that chlorophyll is made up of two compounds, "one green, the other yellow", Flammarion posited that "the yellow chlorophyll can exist alone, or be dominant" on Mars.

## The moons of Mars

William Herschel, among others, had searched the sky in the vicinity of Mars for any satellites, but found nothing.

At the particularly favourable opposition of 1877, Asaph Hall in America tried his hand. A companion would have to be very small, and hence very faint, so to avoid being dazzled by the planet itself Hall positioned it just outside the field of view. He began his search a month prior to opposition. The first few nights were fruitless. After spotting something on 11 August, he had to endure five nights of cloud before being able to observe again, at which time he was delighted to see that it was still present, suggesting that it was moving with the planet rather than being a background star. The proof came the following night when initially it was absent and then emerged from the glare as it travelled around the planet.

To Hall's amazement, on 17 August he discovered a second satellite orbiting much closer in.

Further observation established that the two moons travel in almost circular orbits that coincide with the plane of the planet's equator.

As Mars receded after opposition, its satellites were lost to view. They were not seen again until 1879 and then again in 1881.

Hall named the satellites after the two

---

## THE NEBULAR HYPOTHESIS

Impressed by the rings of Saturn, the French mathematician Pierre-Simon Laplace proposed in 1796 that the solar system was formed by the gravitational collapse of a cloud of gas that was in a state of rotation. This analysis was prompted by what he considered to be several remarkable facts (as far as was known at that time):

(1) All the planets travel around the Sun in the same direction.
(2) All the planets rotate on their axes in that direction.
(3) Apart from minor departures, the planets all travel around the Sun in the same plane.

Laplace said that as the cloud shrank the conservation of angular momentum would have caused its rate of rotation to increase. It would have repeatedly shed material in order to relieve itself of 'excess' angular momentum, thereby making a series of concentric rings in a single plane. After the central mass had become the Sun, each ring would have condensed to create a planet in a near-circular orbit at that particular distance from the Sun. By this reasoning, Mars formed earlier than Earth. In this scheme, Mars was an ancient, eroded world that had rapidly lost its internal heat by virtue of being smaller.

Although this nebular hypothesis was widely accepted at the time, additional mathematical analysis showed it wouldn't work as Laplace had imagined.

---

companions of Mars in Homer's *Iliad*, the inner moon becoming Phobos and the outer one Deimos.

In contrast to the Earth–Moon system, where the separation is of the order of 385,000km and the orbital period is a month, the two Martian satellites are much closer in (this perhaps explained why nobody had seen them earlier; people had been searching farther out).

Phobos orbits at an altitude of only about 6,000km with a period of about 7hr 40min. It travels so rapidly that to an observer on the planet, the moon would rise in the west, race across the sky and set in the east several times per sol. Viewed from Phobos, the disk of Mars would cover almost a quarter of the sky. Being so close to the planet, the polar regions of Mars would not be visible to an observer on the moon.

In contrast, orbiting at an altitude of 20,000km with a period of 30hr 18min, Deimos would rise in the east and travel so slowly as to take 130hr to complete one circuit of the sky.

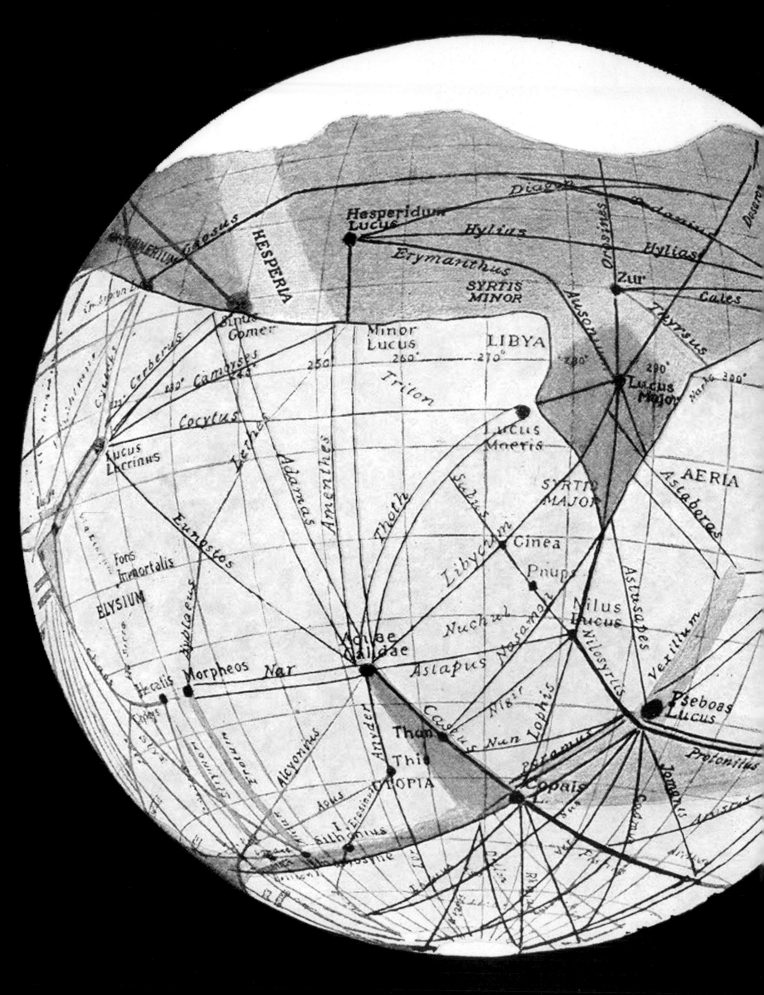

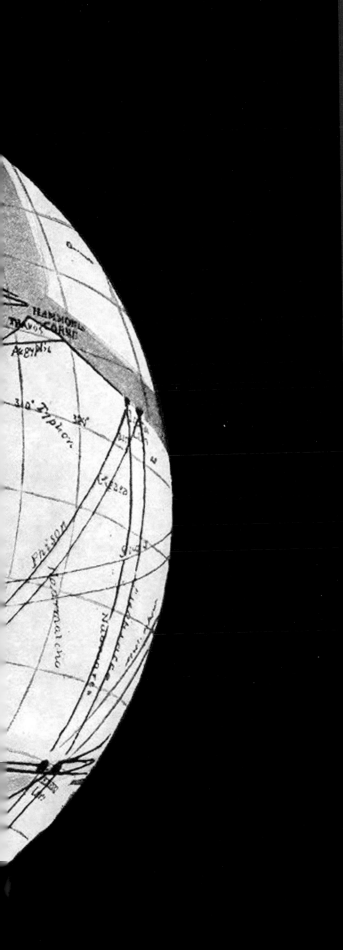

P. L.

# Chapter Two

# Theorising about Mars

In the late 18th century some observers reported long, narrow linear features on Mars, joining up the dark patches. This prompted speculation that the planet was inhabited by an ancient race of intelligent beings who created canals to transport water from the polar ice caps into the arid deserts. This idea was rejected as we learned more of conditions on the planet.

OPPOSITE A depiction of the canals of Mars by Percival Lowell in his 1908 book *Mars as the Abode of Life*.

RIGHT Giovanni
Virginio Schiaparelli
depicted at the
Brera Observatory
in Turin by Achille
Beltrame for the local
*Sunday Courier*,
28 October 1900.

IL GRANDE ASTRONOMO SCHIAPARELLI NELL'OSSERVATORIO DI BRERA.
*(Disegno di A. Beltrame, da ) 1910).*

## Schiaparelli's canali

**G**iovanni Virginio Schiaparelli in Milan carried out a trigonometric survey of Mars during its perihelic opposition of 1877. He installed a micrometer in the eyepiece of his telescope to measure the positions of albedo features relative to a grid of reference points. Because on this occasion the southern hemisphere was best presented, his mapping was primarily in that hemisphere. He introduced a new nomenclature inspired by terrestrial geography and classical literature, with the prominent dark V-shaped feature sketched by Huygens becoming Syrtis Major.

Unlike his predecessors, Schiaparelli sketched the dark areas as possessing sharp, rather than fuzzy boundaries. But the most astonishing aspect of his map was the many

RIGHT A map of the
southern hemisphere
of Mars produced
by Schiaparelli after
observing the planet at
the opposition of 1877,
when the tilt of the
planet's axis favoured
the southern pole.
*(RAS)*

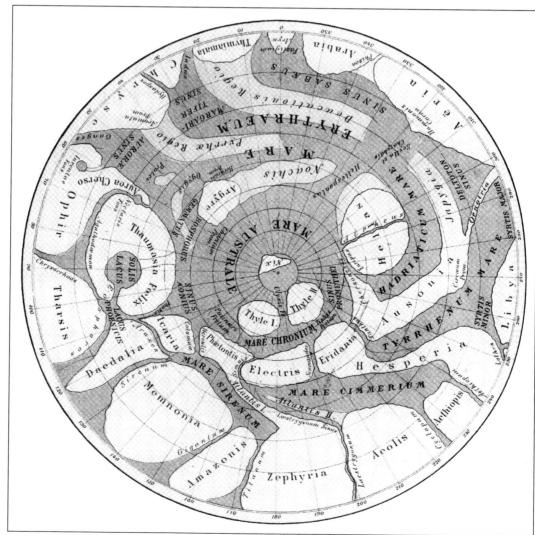

narrow lines that he called *canali*, meaning channels. "The canali run from one to another of the dark areas, usually called seas, and form a well-marked network over the bright part of the surface. Their arrangement appears to be constant and permanent. They traverse the planet for long distances in regular lines that do not at all resemble the winding courses of our streams. Some of the shorter ones do not attain 500 kilometres, but others extend for thousands of kilometres. Their number could not be estimated at less than sixty. Some of the lines are easy to see, others are extremely difficult, and resemble the finest thread of a spider's web drawn across the disk."

Several of Schiaparelli's canali had been recorded by previous observers. Dawes had drawn a few as ill-defined streaks, and there were candidates on sketches by Beer and Mädler, Secchi, Kaiser, and Lockyer. But none of those observers had deemed the features to be noteworthy.

To refine Proctor's map, in 1877 Nathaniel Everett Green observed from the Atlantic island of Madeira, where what astronomers refer to as the 'seeing' was excellent. Green criticised Schiaparelli for "turning soft and indefinite pieces of shading into clear, sharp lines".

Making his own comparison of the observations by Schiaparelli and by Green, Thomas William Webb in England opined that, "much may be due to the different mode of viewing the same objects, to the different training of the observers, and to the different principles on which the delineation was undertaken." In particular, "Green, an accomplished master of form and colour, has given a portraiture, the resemblance of which as a whole commends itself to every eye familiar with the original. Schiaparelli, on the other hand, inconvenienced by colour-blindness, but of micrometric vision, commenced by measurement of 62 fundamental points, and carrying on his work with most commendable pertinacity, has plotted a sharply-outlined chart which, whatever may be its fidelity, no one would at first imagine to be intended as a representation of Mars."

In other words, Webb decided, "one had produced a picture, the other a plan". But

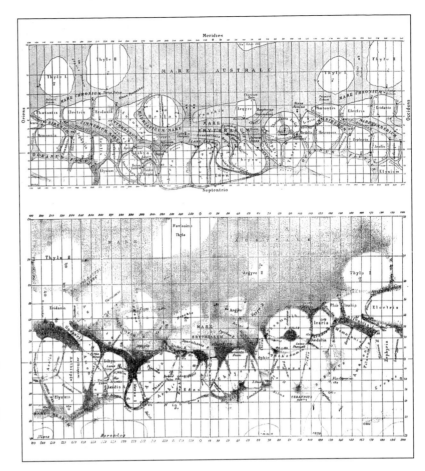

then, to be fair, Green was a professional artist whereas Schiaparelli was an engineer.

In essence, the issue was that sketching the disk of Mars was not an objective process. Each observer saw the planet in his own particular way, according to his eye, his experience, and his method of drawing. Indeed, as Flammarion noted, on one fine evening at the Paris Observatory two observers drew the disk using the same telescope within minutes of one another and there was "little resemblance" between the results.

When Mars returned to opposition in 1879, Schiaparelli confirmed the overall

**ABOVE Maps produced by Schiaparelli in 1877 (top) and 1879. The former spans the latitude range 40°N to 80°S (inverted). Two years later, he was able to extend the northern hemisphere coverage to 60°N.** *(With thanks to Bill Sheehan)*

## A TRANSIT OF EARTH

On the day that Mars reached opposition in 1879, an observer on that planet would have seen Earth cross the disk of the Sun. Although the planes in which Earth and Mars orbit the Sun are slightly inclined, on this occasion the planets were in a line. An observer would also have seen the Moon leading its primary across the solar disk.

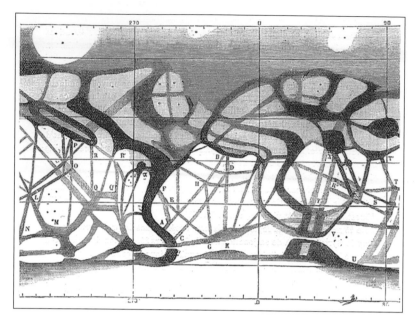

ABOVE **In 1886 Henri Perrotin and Louis Thollon produced this map, which they said confirmed Schiaparelli's discovery of linear features.**

*(With thanks to Bill Sheehan)*

BELOW **A map by Schiaparelli that integrated his observations between 1877 and 1886.**

now showed a plethora of fine detail. In retrospect, it seems that when Mars was at perihelion two years earlier it had been partially obscured by dust storms.

The opposition of 1881–82 was not so favourable because Mars did not come as close, but the seeing in northern Italy was excellent and Schiaparelli continued his study noting the appearance and disappearance, often over the space of only a few days, of further geminations. Accepting that the style of his 1877 map was "purely schematic", Schiaparelli set out this time to make a map which was "more pleasing to the eye".

## Other observers

For almost a decade, Schiaparelli remained the only person to claim to have seen the canali as narrow lines but in April 1886 Henri Perrotin and his assistant Louis Thollon at the Nice Observatory confirmed their existence, saying that they were "in nearly all respects, almost the character attributed to them [by Schiaparelli]".

Despite that year's opposition not being very favourable, additional supportive reports came from England and the United States. One eager observer, François Joseph Charles Terby in Belgium, inspected the planet "map in hand"

character of the canali and announced that a prominent one had 'geminated' (as he put it) into a pair of lines which ran in parallel, a short distance apart. He also said that some areas which he had drawn as featureless

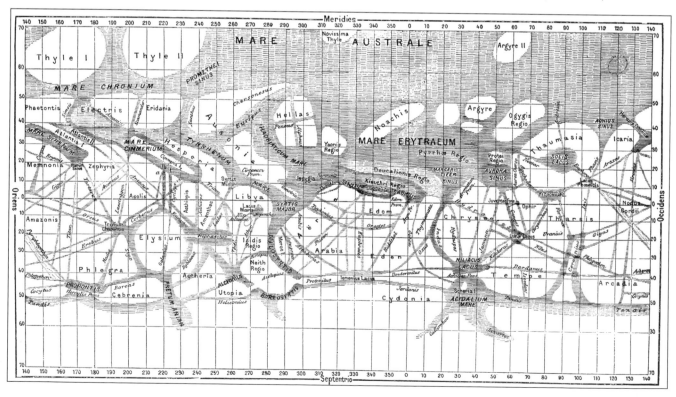

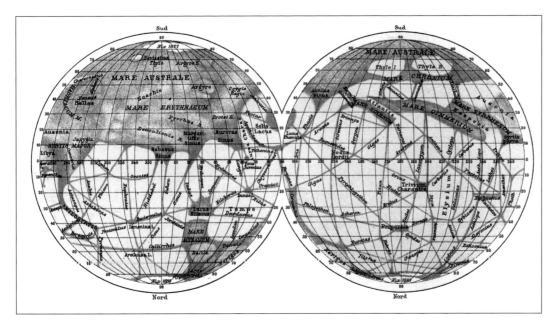

LEFT This two-hemisphere perspective summarised Schiaparelli's findings between 1877 and 1888.

to ensure that he knew where to look. But many other observers using fine telescopes had no luck.

Nevertheless, people were eager for the next perihelic opposition in order to get a really good look at them.

Although the 1888 opposition was near-aphelic, Schiaparelli, observing with a large new telescope, argued the canali retained "the distinctness of an engraving". To wrap up his study of the planet, that year he integrated his results as a map of the entire globe.

In the meantime, there was a significant advance in telescope development. In 1856 the Italian astronomer Charles Piazzi Smyth had temporarily erected a telescope on the 12,200ft summit of the volcanic island of Tenerife in the Atlantic and reported excellent seeing. In 1874, James Lick, having gained a real-estate fortune in the Californian Gold Rush of 1849, gave $700,000 for the construction of a telescope with a lens diameter of 36in, making it the world's largest at that time. It was commissioned in January 1888 on the summit of the 4,250ft Mount Hamilton in California and soon became a major factor in advancing astronomy.

The near-perihelic 1892 opposition of Mars was too far south in the sky for a serious study by northern observers but William Henry Pickering had a good view from Arequipa in the Peruvian Andes. He drew the canali as hazy streaks, and in exceptional

seeing was able to resolve small dark spots where canali intersected. As the canali had hitherto been seen only on the ochre areas, Pickering was astonished to find a faint line crossing one of the dark areas. Rejecting open water, Pickering opted for the vegetative hypothesis proposed by Liais 30 years earlier. He suggested that the canali were swathes of vegetation living off volcanic gases that leaked from deep cracks in the crust.

BELOW Constructed in the 1880s on Mount Hamilton in California, the Lick Observatory was the world's first high-altitude astronomical facility. This picture was taken circa 1900.
(Lick Observatory)

## Lowell's Martians

Upon graduating from Harvard in 1876, Percival Lowell was appointed to run one of the family businesses. But he had a passion for astronomy, and in 1893, after reading Flammarion's books and corresponding with Pickering, he decided to set up an observatory "to investigate the conditions for life on Mars" at the perihelic opposition of 1894. He confidently stated, "there is strong reason to believe that we are on the eve of pretty definite discovery in the matter".

With Pickering's assistance, Lowell sited his observatory at Flagstaff, a small railroad halt in what was then the Arizona Territory which, by virtue of being on a plateau at an elevation of 7,200ft, had excellent seeing for a large fraction of the year. Although time was short, a borrowed refractor with a lens diameter of 18in was ready for 'first light' on 23 April 1894.

Beginning on 28 May, Lowell, Pickering, and their assistant Andrew Ellicott Douglass observed Mars on virtually every night through the opposition and on into the following April, compiling almost 1,000 drawings. As Lowell delightfully wrote, the canali were visible "hour after hour, day after day, month after month".

Lowell was unperturbed by the fact that he could not find all of Schiaparelli's canali, because his predecessor had said that their appearance evolved during the observing season and from one opposition to the next. In fact, Lowell recorded many more of them than his mentor.

In addition to mapping the small dark spots where canali intersected on the ochre tracts, Douglass verified Pickering's hitherto unconfirmed report of canali running across dark areas.

On returning to Boston, Lowell mulled over his observations. He agreed with Pickering that the canali were swathes of vegetation but rejected the idea that this formed at crustal fractures. Instead, Lowell believed that the canali were laid out purposefully.

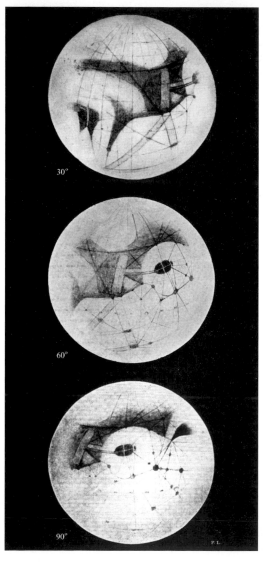

**RIGHT** Having accumulated almost 1,000 drawings of Mars during the 1894 opposition, Lowell wrote a book entitled simply *Mars*. Published late the following year, it featured a series of hemispheric views with the planet rotating 30° axially each time. In this way he provided a global perspective of the system of canals, painstakingly compiled from the many observing sketches. Three of his hemispheric views are shown here. Note that because south is at the top, the rotation seems to be in the wrong direction. *(Lowell Observatory)*

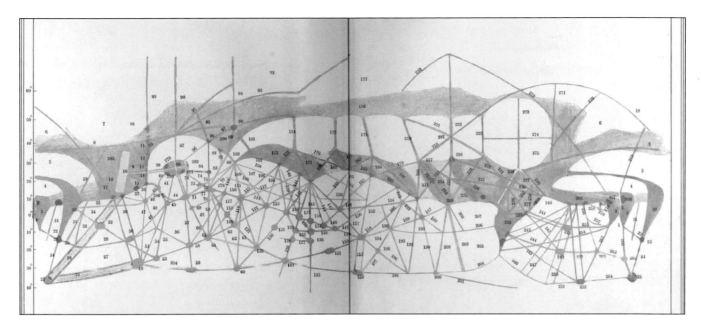

Lowell published his conclusions in late 1895 in a tome entitled *Mars*: "Firstly, that the broad physical conditions of the planet are not antagonistic to some form of life; secondly, that there is an apparent dearth of water on the planet's surface and, therefore, if beings of sufficient intelligence inhabit it, they would have to resort to irrigation to support life; thirdly, that there turns out to be a network of markings covering the disk precisely counterparting what a system of irrigation would look like; fourthly, and lastly, there is a set of spots placed where we should expect to find the land thus artificially fertilised, and behaving as such constructed oases should."

Hence he inferred that the Martians lived where the canali intersected on the ochre tracts, at the dark spots which he called oases.

Lowell's portrayal of Mars as a dying world was so evocative that it inspired Englishman Herbert George Wells to write a story called *The War of the Worlds*, in which the Martians invade Earth. After serialisation in 1897 it was issued as a book in 1898 and became an instant best-seller.

Although the public was fascinated by Lowell's picture of Mars, many highly skilled observers with excellent telescopes were still finding the canali elusive.

## Critics

Having observed Mars in 1877 using a refractor comparable to that employed by Schiaparelli, Henry Pratt in England stated that in exceptional seeing conditions "glimpses were obtained of a structure so complicated and delicate that the pencil cannot reproduce it". In likening this to stippling, he continued, "What at first sight appeared as a broad hazy streak [could be briefly] resolved into several separate masses of shading enclosing lighter portions full of very delicate markings."

At the opposition of 1892 Charles Augustus Young noticed that whenever he saw a hint of a canal using a refractor similar to those of Schiaparelli and Pratt, it was shown to be illusory by a much larger such instrument.

That same year, Edward Emerson Barnard wrote of a small dark spot on the ochre tract: "It is connected with the great sea south by a slender thread-like line. There is a small canal running north from Solis Lacus to a diffused dusky spot that does not appear on Schiaparelli's chart." But in 1896 he stated that although the surface of Mars was "wonderfully full of detail … to save my soul I cannot believe in the canals as Schiaparelli and Lowell draw them. I see details where they have drawn none. [And] I see details where some of their canals are, but these are not straight lines at all. When best seen, these canals are very irregular and broken up."

When seeing was exceptional, Barnard noted a plethora of small faint marks that were so intricate as to be impracticable to sketch. Some of the irregular dark streaks matched the tracks of some of Lowell's well-defined canali.

**ABOVE** Lowell's **1895 book *Mars* also included a map showing the canals.** *(Lowell Observatory)*

Significantly, he also saw irregular fine detail on the dark areas, supporting the growing belief that these were not shallow seas. Despite positive sightings of canali by people using small telescopes, Barnard opined that "before many oppositions are past" they would be established as "a fallacy".

Having studied Mars using a variety of telescopes, Pickering was now of the belief that the linear separation of the twin canals "was inversely proportional to the diameter of the telescope used and directly proportional to the distance of the planet. In other words, if we use a telescope of twice the diameter we shall find the same canals will measure only half as many miles apart." Hence, "while the canals are undoubtedly genuine, their doubling is an optical illusion".

Vincenzo Cerulli in Italy observed Mars in 1897. When one prominent streak "lost its form of a line and altered itself into a complex indecipherable system of tiny patches" he, too, decided canali were illusory. Cerulli was later fascinated to discover that observing the Moon through low-power opera glasses suggested canali-like streaks which he knew were not real structures; they were fine detail that his sense of perception was linking up.

Edward Walter Maunder in England came to a similar conclusion in 1903. Accepting that Schiaparelli drew what he had seen, Maunder argued that what Schiaparelli saw as a thin straight line was an interpretation of a mass of detail just beyond the resolving power of his telescope in those seeing conditions.

Maunder arranged with Joseph Edward Evans, a friend who was headmaster of a school in Greenwich, for a number of schoolboys to perform an experiment. The boys, who had no knowledge of the purpose of the test, were positioned at different distances from a disk bearing an impression of the light and dark areas of Mars, augmented by a plethora of fine dots. The nearest boys noted the dots and drew them distinctly. Boys farther out saw only the general albedo features. Significantly, boys on the limit of resolution for the dots drew fine lines. Lowell of course rejected what he referred to as "the small boy hypothesis".

In 1894 Lowell had observed a broad blue band develop at the periphery of the southern polar cap as that retreated. First reported by Beer and Mädler, this feature had been rejected in 1892 by John Martin Schaeberle at Lick as just an illusion induced by the contrast between the brilliant white cap and the adjacent ochre tract. Lowell, however, was sure it was real.

When Arthur Cowper Ranyard and George Johnstone Stoney in England argued that the caps were a frost of frozen carbon dioxide, Lowell insisted: "At pressures anything like one atmosphere, carbon dioxide passes at once from the solid to the gaseous state. Water, on the other hand, lingers in the intermediate stage of a liquid."

But pressure was the issue. Unable to measure the pressure at the surface directly, Lowell reasoned inversely: if the band was aqueous and the melt-water was regenerating vegetation then this required the atmosphere to be sufficiently dense to maintain a moderate temperature. Despite the vast extent of the south polar cap in winter, the rate at which it contracted meant the ice was so thin that there was no more water on the planet than was present in the Great Lakes of America.

## Lowell undeterred

Lowell installed a 24in refractor at Flagstaff in 1896 and continued to study Mars at successive oppositions, but with the range increasing it was difficult to improve on his initial observations. Nevertheless in 1906 he followed up his first book with the more provocative *Mars and its Canals*, in which he painted a more complete picture of the planet's inhabitants. The planet was being overwhelmed by a rapid process of desertification and the polar caps were the only remaining sources of water. In a valiant effort to sustain their civilisation the inhabitants had excavated the canals to transport the melt-water to the equatorial zones.

Lowell boldly stated, "That Mars is inhabited by beings of some sort we may consider as certain as it is uncertain what these beings may be." He offered no opinion as to their physical form, but developed a clear sense of empathy. The "cosmopolitan" scope of the canal network indicated a unified society. "Girdling the globe and stretching from pole to pole, the Martian canal system not only embraces their whole world, but is an organised entity. Each canal joins another, which in turn connects with a third, and

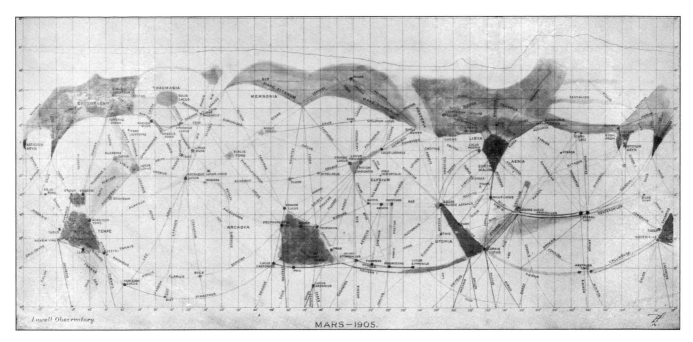

MARS—1905.

so on over the entire surface of the planet. This continuity of construction posits a community of interest. … The thing that is forced on us in conclusion is the necessary intelligent and non-bellicose character of the community which could thus act as a unit throughout its globe."

Specifically, Lowell argued that the Martians, faced with the extinction of their species, had banished warfare. "War is a survival among us from savage times and affects now chiefly the unthinking elements of the nation. The wisest realise that there are better ways for practising heroism, and other, more certain ends of ensuring survival of the fittest." (Less than a decade later humans launched their first 'world war'.)

It was no accident that the canali were interpreted as artificial waterways. On Earth canals were the state of the art in transportation systems. The Suez Canal had been completed in 1869, and the Panama Canal was under construction. It seemed reasonable that the more advanced and more motivated inhabitants of Mars should have been able to build canals on a global scale.

Indeed, in *The Riddle of Mars*, published in 1914, Charles Edward Housden went so far as to explore in detail the engineering issues that the Martians must have overcome.

## Lowell rebutted

Alfred Russel Wallace, the English naturalist whose research reinforced Charles Darwin's

*On the Origin of Species*, and who published his own *Contributions to the Theory of Natural Selection* in 1870, was asked to review Lowell's *Mars and its Canals*. Wallace was so appalled that in 1907 he published his rebuttal in the form of a book entitled *Is Mars Habitable?*

In particular, Wallace took issue with Lowell's opinion that the temperature at the surface of Mars compared favourably with that of a summer's day in the south of England. Wallace argued that Mars must be sub-zero, and hence

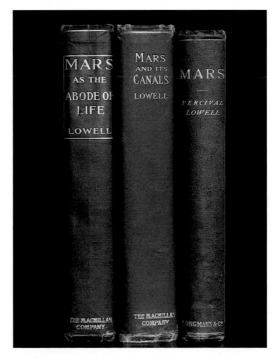

**ABOVE** On 27 August 1911 *The New York Times* praised Lowell's claim of life on Mars.

water could not possibly flow across the surface in open channels. "All physicists," he wrote in summary, "are agreed that, owing to the distance of Mars from the Sun, it would have a mean temperature of about −35°F even if it had an atmosphere as dense as ours. But the very low temperatures on Earth at the equator, at a height where the barometer stands at about three times as high as on Mars, proves, that from the scantiness of atmosphere alone Mars cannot possibly have a temperature as high as the freezing point of water," which, he insisted, was "wholly incompatible with the existence of animal life". Wallace's conclusion was that Mars "is not only uninhabited by intelligent beings such as Mr. Lowell postulates, but is absolutely *uninhabitable*". In fact, Wallace erred in calculating the temperature on Mars but his analysis was still superior and his conclusion stands.

Eugene Michael Antoniadi observed Mars during its 1909 perihelic opposition using a refractor at Meudon, near Paris, which was only slightly smaller than that at Lick. His

**RIGHT** In 1909 Eugene Michael Antoniadi produced sketches during times of excellent 'seeing' which meant that what some observers saw as linear features were really a case of the human eye perceiving order from a chaos of fine detail. In his opinion, the canals were an optical illusion.

**BELOW** Supporters of Lowell argued that the linear features on Mars were fertile tracts spanning the artificial watercourses on otherwise arid deserts, much like the River Nile in Egypt. Nowadays we can view the Nile from orbit. This picture was taken by astronaut Douglas H. Wheelock aboard the International Space Station in 2010. *(NASA)*

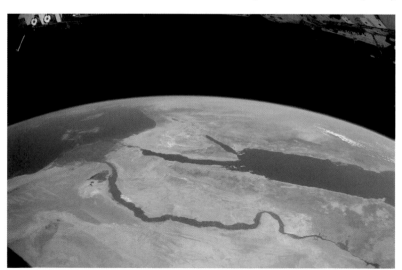

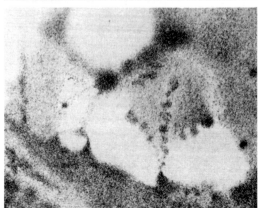

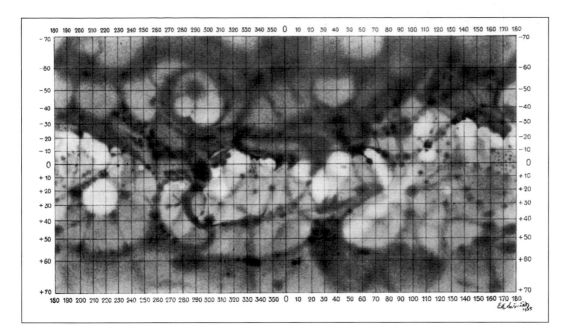

LEFT A map of Mars made by Antoniadi incorporating observations made between 1905 and 1929.

BELOW A map by Gerard de Vaucouleurs in the early 1940s to illustrate a thesis in French about Mars. It was later translated into English by Patrick Moore and published in 1950 as *The Planet Mars*.

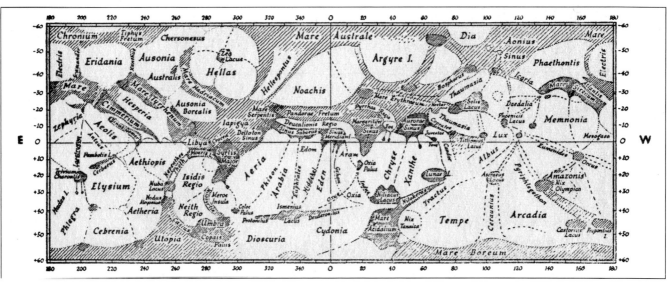

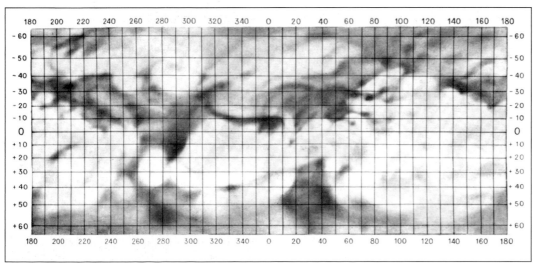

LEFT In 1957 the International Astronomical Union issued this 'official' map of Mars based on observations by Antoniadi. *(IAU)*

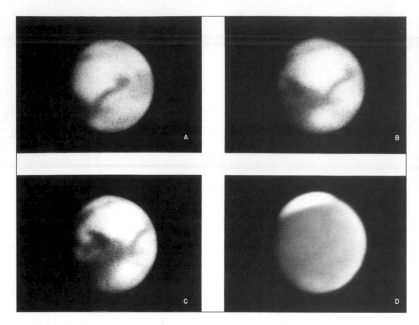

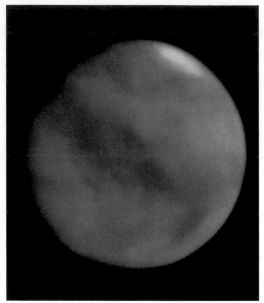

**ABOVE** The axial rotation of Mars is illustrated by these pictures taken by the 100in Hooker reflector on Mount Wilson in California. *(Mount Wilson Observatory)*

**BELOW** Looking to the future, in 1962 the Aeronautical Chart and Information Center of the US Air Force issued this map of Mars, based on work by Vesto Melvin Slipher of the Lowell Observatory. *(USAF/NASA-Lunar and Planetary Institute)*

**ABOVE** Although a photograph could not capture the clarity that was evident to a visual observer in instants of good 'seeing', this picture of Mars, taken during the particularly favourable 1956 opposition by the 60in reflector atop Mount Wilson, was one of best prior to the Space Age. *(Mount Wilson Observatory)*

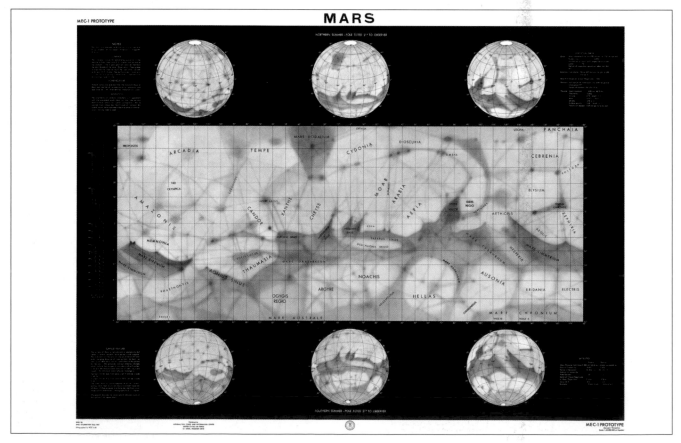

first inspection on 20 September was during excellent seeing and he was astonished at the sight. "I thought I was dreaming," he wrote to Lowell. "The planet revealed a prodigious and bewildering amount of sharp or diffused natural irregular detail, all held steadily, and it was at once obvious that the geometrical network of canals Schiaparelli discovered was a great illusion." Antoniadi was a skilled sketcher but what he saw on the disk "could not be drawn, hence only its coarser markings were recorded in the notebook".

When Lowell died in 1916 the public was still intrigued by his portrait of Mars as home to an ancient race of intelligent beings, but the scientific community had rejected him as a well-meaning crank.

Lowell and Wallace had both based their estimates of the surface temperature of Mars on indirect reasoning. It was not until the invention of the thermocouple in the early 1920s that it became possible to measure such a thing directly.

In 1926, William Weber Coblentz and Carl Otto Lampland at Flagstaff found that when the planet presented a gibbous phase they could track a site over the terminator into darkness and measure the rate at which the temperature fell. An extrapolation implied the surface must plunge below –75°C (possibly –100°C) at local midnight. Such a rapid radiation of heat to space indicated that the Martian atmosphere was very thin. Even though the large dark areas were typically

**BELOW A map originally prepared by Antoniadi was redrawn by Lowell Hess for the book _Exploring Mars_, written by Roy A. Gallant. It in turn was projected onto a globe by Tom Ruen. Although the albedo features match up with imagery by the Hubble Space Telescope, the system of canals simply doesn't exist.** _(Tom Ruen/ Eugene Antoniadi/Lowell Hess/Roy A. Gallant/ NASA/STScI)_

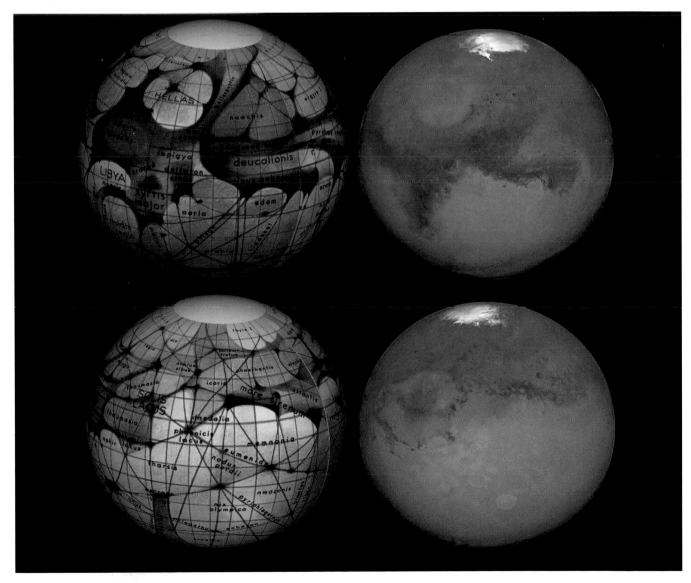

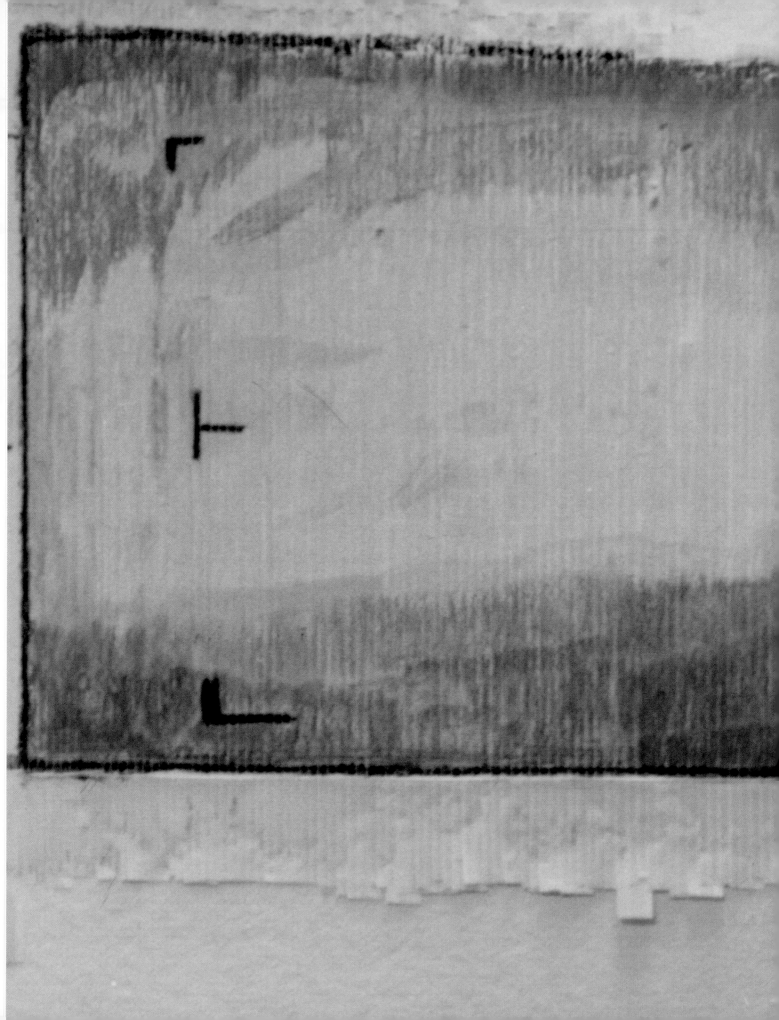

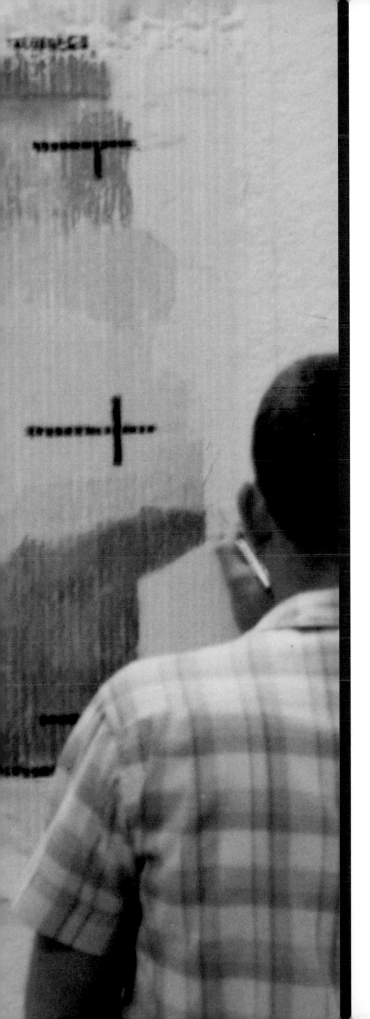

## Chapter Three

# Mars shock

─────●─────

Sending probes through space to inspect Mars directly, initially during flybys and later from orbit, showed it to be very different to expectation. It was nothing like a small version of Earth. Its ancient surface looked so desolate that the planet lost much of its popular appeal. Nevertheless, it was a paradise for scientists with an interest in comparative planetology.

**OPPOSITE** When the data began to stream in from Mariner 4, eager engineers assembled the first image by manually colouring in the pixels onto a grid pattern. *(NASA/JPL-Caltech)*

# Flyby snapshots

In 1925 the German mathematician Walter Hohmann calculated trajectories from Earth to Mars and discovered that the minimum energy 'transfer' would depart on an elliptical path which was tangential to Earth's orbit around the Sun and had an aphelion tangential to the orbit of Mars. The passage would take about 250 days. In that time Mars would travel an arc of 130° around its orbit. The optimal 'launch windows' opened shortly prior to opposition and lasted several weeks. Hohmann identified several other trajectories that required greater expenditures of energy.

Of course, this scheme depended on the capability of the rocket. Since Mars has an elliptical orbit whose closest point of approach to Earth varies greatly, the minimum being about 56 million km, it would be best to attempt such a voyage at a perihelic opposition.

## Early frustrations

Unfortunately the technology for travelling in space did not exist at the very close perihelic opposition of 1956, but having begun the Space Age on 4 October 1957 by launching Sputnik, the Soviet Union decided to dispatch two probes during the October 1960 window. The plan was to insert a probe into a 'parking' orbit around Earth as a preliminary to firing a small 'escape' stage to achieve an interplanetary trajectory. Unfortunately, the launch vehicles malfunctioned and the probes failed to reach parking orbit. Two more probes were prepared for the 1962 window. The first was stranded in Earth orbit by the failure of its escape stage, but the second, launched on 1 November, succeeded and was announced as Mars 1. Although it suffered attitude control problems and its radio fell silent after several months, the fact that it passed within 200,000km of its objective in June 1963 made it a great achievement.

Meanwhile in America, the National Aeronautics and Space Administration (NASA) had attempted two interplanetary missions to investigate Venus. The first was lost at launch but Mariner 2 became the first to provide close observations of another planet when it made a 35,000km flyby of its target on 14 December 1962.

Although the opposition of Mars in March 1965 was aphelic and the Hohmann transfer was not favourable, missions were still attempted. The Soviets had a new design of interplanetary spacecraft. They launched Zond 2 on 30 November 1964. It fell silent in transit and flew past the planet on 6 August at a range estimated to have been as much as 650,000km.

## The first success

Although NASA lost one probe at launch in the 1964 window for Mars it was able to undertake modifications and dispatch Mariner 4 on 28

## THE MARINER 4 SPACECRAFT

The basic structure of the vehicle was an octagonal frame.

A cruciform of fold-down solar panels with a total of 28,224 solar transducers were to provide 300W of electrical power during the flyby of Mars.

Because the vehicle was required to be 3-axis stabilised in space, the attitude control system could tilt small vanes installed on the tips of the panels to correct any imbalanced forces imparted by the 'solar wind', a plasma emanating from the Sun that streams through interplanetary space at high speed.

The elliptical dish of the high-gain antenna was affixed on the top of the frame in an orientation such that Earth would be in its narrow beam during the encounter with Mars. On the tip of a mast alongside the dish was the low-gain antenna which would receive commands from Earth.

Seven of the enclosed bays of the frame contained spacecraft and scientific systems; the other bay held the liquid-propellant rocket motor that would perform the mid-course correction.

In addition to the imaging system for use during the flyby of Mars there were six 'particles and fields' experiments to study the solar wind while in transit to the planet and in its immediate vicinity, in particular to determine whether Mars has a magnetic field.

In its deployed configuration the spacecraft spanned 6.8m and was 3.3m tall. Scientific instruments and associated systems accounted for 10% of its mass of 260kg.

OPPOSITE **Details of Mariner 4.** (NASA/JPL-Caltech/Woods)

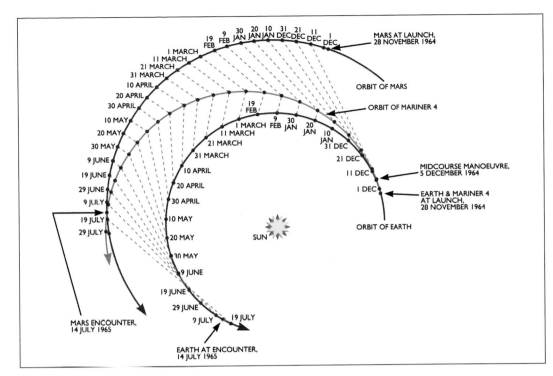

November, a few days before that year's window closed.

There were several rules for the trajectory during the brief encounter. Firstly, neither Mars nor its two moonlets could be permitted to obstruct the spacecraft's view of the star by which the vehicle orientated itself; nor could it fly through the shadows of any of these bodies. As the 228-day trajectory was designed to catch up with the planet, the observations would be of the trailing hemisphere. The post-encounter trajectory had to pass behind the planet (as viewed from Earth) to allow the manner in which the radio signal was cut off by the limb of the planet to yield the first definitive measurement of the density of the atmosphere. A second such measurement would be obtained when the craft emerged from the leading limb. Crucially, this occultation had to occur when the planet was above the horizon of the receiving antenna at the Goldstone station in California.

A mid-course correction on 5 December

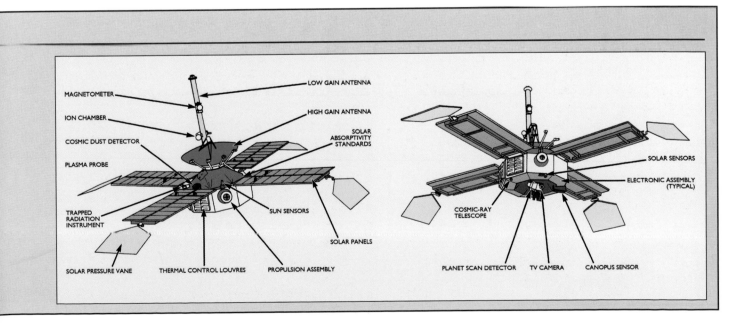

trimmed the initial 240,000km 'miss distance' to the intended 10,000km flyby. In addition to arranging the timing, this manoeuvre determined the surface features that would be in the field of view of the camera at the time of closest approach. The objective was to obtain pictures comparable in resolution to contemporary telescopic pictures of the Moon. It had initially been intended to inspect the prominent dark feature Syrtis Major, but the delayed launch ruled this out.

On the octagonal base of Mariner 4 was a vidicon imaging system carried on a platform which could be rotated through 180°. The system was powered up 6hr prior to the encounter on 14 July 1965 to allow time to troubleshoot any problems – diagnosing a fault and issuing recovery actions needed to be prompt, since the light-speed travel time over such a distance was 12min. An hour out, the platform rotated until its wide-angle sensor noted the presence of the planet, at which time it centred the camera on the target. The imaging sequence was started when the narrow-angle sensor detected the planet about half an hour later. The image from an f/8 Cassegrain telescope that had a 4cm aperture and a 30cm

focal length was projected onto a phosphor screen to be digitised into an array of 200 x 200 pixels, each encoding a 6-bit greyscale value.

The data from the particles and fields instruments was transmitted in real time but the images were stored on a 100m loop of magnetic tape. The exposure time had been pre-set at 1/20th of a second; this being the best estimate on the basis of predicted illumination. It took 24sec to read out the image from the vidicon and another 24sec to clear the screen, hence pictures could be taken no more rapidly than once every 48sec. A rotating shutter alternated blue-green and orange-red filters to enhance the contrast in the greyscale and also to emphasise the colour differences of the albedo features on the planet. About 12hr after the encounter, the tape began to replay the images. At 8.33bit/sec using the 10W transmitter, it required 8hr 20min to replay the 240,000 bits in each image. In fact the effective rate was one frame every 10hr, because each frame was accompanied by some engineering data.

## Not what was expected!

Taken from a range of 16,500km, the first frame showed the planet's limb and the blackness of space. The engineers were ecstatic, because this confirmed that the camera had properly locked onto the planet. As the other frames were processed there was disappointment. Despite having used filters to highlight the contrast on the planet's surface, most of the 22 images were bland due to 'flare' in the optics, and the remainder were black.

Fortunately, the Jet Propulsion Laboratory (JPL) of the California Institute of Technology in Pasadena, which operated deep space missions for NASA, had a computer algorithm to 'enhance' digital imagery in order to 'stretch' the contrast, and when this was applied some surface details became evident.

The imaging sequence began with a view looking northwards across the limb at a longitude of 190° (on Mars longitudes are west of the meridian) and tracked southeast and across the equator at longitude 180° to about 52° south, where it swung north and crossed the terminator. In all, it spanned some 90° of longitude.

In the first few frames, the surface appeared to be blotchy with hints of large circular features. On frame #7 it became apparent that

**BELOW** A track of the images taken by Mariner 4 projected on the best map of Mars available at that time. Note that unlike astronomers, NASA placed north at the top. The sequence started with an oblique view over the local horizon, ran south and east, then over the terminator into darkness. *(NASA/JPL-Caltech)*

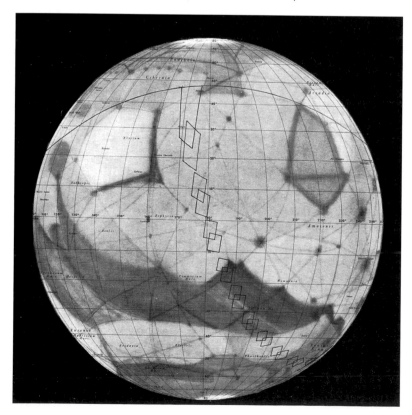

these were craters, which were depicted in greater clarity with each successive frame. The crater, 120km in diameter on frame #11, which was taken from a range of 12,500km, became the iconic image of the mission.

The final frame to show surface detail was #15. Although the contrast on the terminator would have emphasised the surface relief, the sensor to increase the exposure as the field of view darkened failed to operate.

Although the imaging system was marred by flare, it had achieved its primary objective of revealing the nature of the Martian surface in a swathe covering about 1% of the planet.

The fact that craters were present on a variety of albedo features familiar to telescopic observers implied the entire surface of Mars was billions of years old, like that of the Moon.

Around 1950, Clyde William Tombaugh had considered the possibility of Mars having suffered impacts that would make big craters. He imagined that the linear streaks that appeared to radiate from the features that Lowell called oases might be deep cracks in the crust produced by the shock of the impacts that excavated the nodes. Ralph Belknap Baldwin and Ernst Julius Öpik independently came up

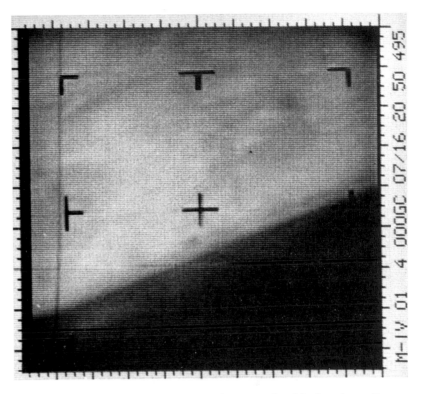

**ABOVE** The first image in the Mariner 4 sequence (upside down) was the local horizon. *(NASA/JPL-Caltech)*

**BELOW** This is the manually assembled version of the first image of Mars from Mariner 4. It is on display at JPL. *(NASA/JPL-Caltech)*

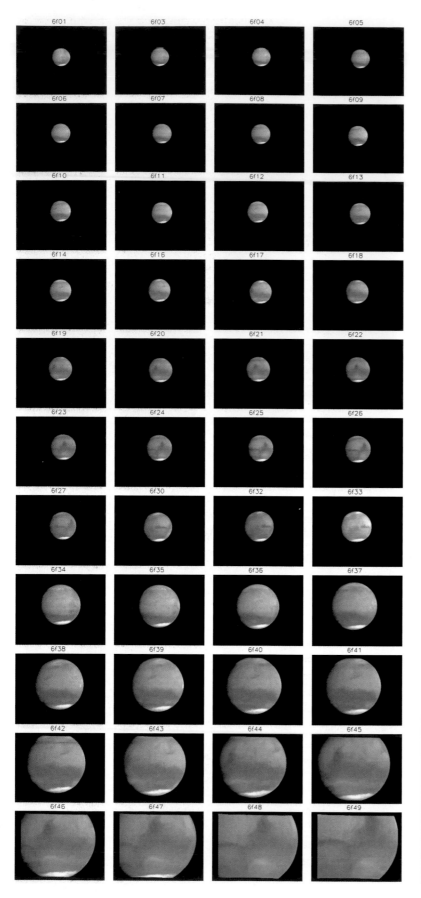

**The far-encounter imaging sequence by Mariner 6 as it closed in on Mars provided useful data for comparison with astronomical maps.** *(NASA/JPL-Caltech)*

and raised rims, but the larger ones appear 'shallower' than their lunar counterparts, having complex central peaks, terraced inner walls, degraded rims and ejecta blankets, and floors that have been levelled by various infill materials. There was a striking paucity of chains of secondary craters made by ejecta from major impacts.

**BELOW A selection of flyby images from Mariner 6.** *(NASA/JPL-Caltech)*

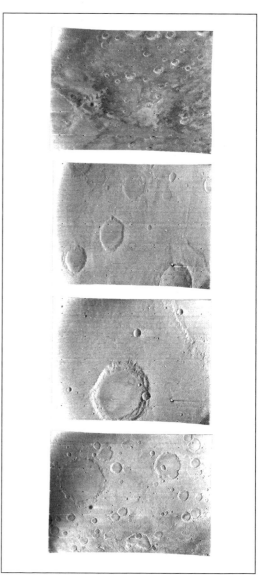

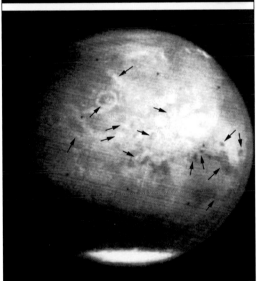

**RIGHT** A track of the images taken by Mariner 7 projected on the best map of Mars available at that time. *(NASA/JPL-Caltech)*

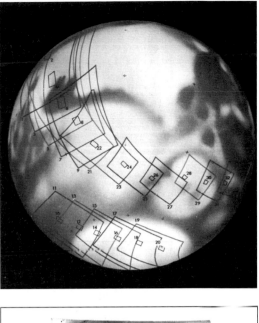

**LEFT** An image of Mars taken by Mariner 7 from a range of 320,000km. A comparison with an image taken by the Hubble Space Telescope in 2003 shows that a lot of real detail was present in the far-encounter image. Astronomers promptly recognised the large circular feature as Nix Olympica, which they had believed to be a high plateau that appeared bright in telescopes because it attracted snow, but now it looked as if it might be a vast crater. Unfortunately, to understand a low-resolution image it is often necessary to have knowledge of the true situation. So the nature of Nix Olympica and most of the other albedo detail would not be understood until the planet was mapped from orbit. *(NASA/JPL-Caltech/STScI/Harland)*

**RIGHT** A selection of flyby images from the southern part of the Mariner 7 track. *(NASA/JPL-Caltech)*

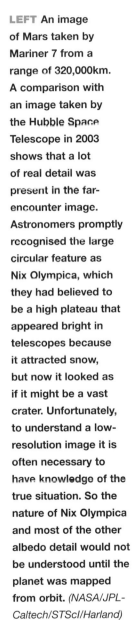

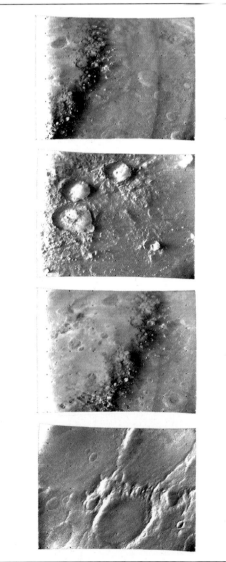

A significant result of the 1969 flybys was the realisation that craters are not ubiquitous. Some areas were classified as 'chaotic' and others as 'featureless'.

Some assumptions of telescopic observers proved to be wide of the mark. A bright circular area called Hellas that had often brightened to resemble an offset polar cap had been interpreted as a plateau that attracted a covering of snow in winter. The pressure data revealed it to be a very deep cavity. It was inferred that when its floor was masked by clouds, it appeared bright and featureless. Another intriguing feature in the far-encounter imagery was a large circular structure where Schiaparelli had often mapped a bright spot that he named Nix Olympica (Snows of Olympus).

The occultation data from these probes verified that the pressure profile of the atmosphere is shallower because Martian gravity is weaker. The surface at Sinus Meridiani was 6.5 millibars. However, the fact that the southern latitudes were 3.5 millibars implied this was elevated terrain. It was decided to adopt 6.2 millibars as a sort of 'sea level' reference altitude. This pressure was significant for being the 'triple point' of water. In fact, given the geographical and seasonal variations, the mean surface pressure on Mars is near this value. Water cannot exist in its liquid phase at lower pressures.

The temperature at the south pole was measured to be −123°C, reinforcing the argument that the seasonal polar caps were frozen carbon dioxide because this was close to the 'frost point' of carbon dioxide.

Between them, the three NASA flybys documented about 10% of Mars at a moderate resolution. Their data repudiated the soundly

reasoned impression of the Red Planet, turning the presumed cold dry world that harboured vegetation into one that still bore the scars of ancient impacts, had an atmosphere too thin for surface water, and was devoid of life.

# A global perspective

Nevertheless, geoscientists were eager to insert a vehicle into orbit to carry out global mapping.

In fact, it would be better to employ two spacecraft. When we view the Moon at its 'full' phase we see only albedo variations. At other phases, shadows at the terminator indicate surface relief. No telescopic study of Mars had ever hinted at topography on its terminator, which was why observers had focused on mapping albedo features. A craft in orbit would be able to view the surface under a variety of illuminations but whereas a high point over the illuminated hemisphere would favour albedo studies, a low point over the terminator was required to undertake topographic mapping.

Accordingly, NASA assigned Mariner 8 a highly inclined orbit that dipped low over the terminator to map 70% of the planet at high resolution, while Mariner 9, in an almost equatorial orbit at high altitude, would monitor the seasonal albedo variations.

As these new spacecraft would require a rocket engine and the propellant to slow down in order to enter orbit around Mars, they would be much heavier than their predecessors. Fortunately, the perihelic opposition of 1971 would minimise the energy requirements.

When Mariner 8 was lost at launch, the plan was redesigned to allow a single spacecraft to make as many observations as possible from a compromise orbital inclination of 65° which would provide illumination at shallower angles than would be ideal for albedo studies and higher than was ideal for mapping the topography. To everyone's relief the reprogrammed Mariner 9 was successfully dispatched on 30 May 1971.

## Dust storm!

Astronomers had seen 'yellow clouds' sometimes blanking out large areas of the planet's surface. In 1909 Antoniadi had seen one that lasted for several days. In 1911 he saw

**BELOW A dust storm can mask the surface of Mars, as these Hubble Space Telescope images for September 2001 illustrate.** *(NASA/STScI)*

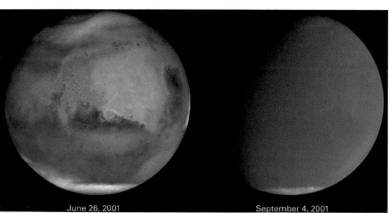

June 26, 2001

September 4, 2001

another that rapidly expanded to mask a large part of the southern hemisphere for several months. Further observations implied they were fine dust stirred up by the strong winds created by differential solar heating in the southern hemisphere at perihelion.

In February 1971, Charles Franklin Capen in Flagstaff predicted that a major dust storm was likely to occur during that year's opposition, and warned that this might interfere with the forthcoming orbital mapping mission.

On 21 September a storm did indeed develop. It was first photographed by Gregory Roberts in South Africa, and by 27 September it hid a wide area to the west of Hellas. By the end of the first week of October the mid-southern latitude zone was masked, and by the end of the month the entire globe was obscured!

The Soviet Union had launched a pair of spacecraft to Mars, each of which released a descent probe 4hr prior to inserting itself into orbit.

Mars 2 entered the intended 1,280 x 24,900km orbit on 27 November but its radio link was so poor that little usable data was received. Its probe penetrated the atmosphere at a steeper angle than planned and struck the surface before it could deploy its parachute.

Despite suffering a propellant leak, on 2 December Mars 3 was able to limp into an orbit far more elliptical than planned, with a very high apoapsis. Thus its opportunities for observing the planet were severely limited, and in any case it was unable to see anything due to the dust storm. The most useful data came from radio occultations.

The Mars 3 probe entered the Martian atmosphere at 5.7km/sec at an entry angle of less than 10°. The drogue parachute was deployed. This drew out the main parachute. The heat shield was jettisoned and the radar switched on. At a height of 20 to 30m and falling at 60 to 110m/sec, the parachute was discarded and a small rocket lifted it away from the lander. Meanwhile, the lander fired its own retrorockets. After an entry and descent phase lasting a little over 3min the lander touched down at a speed of 20.7m/sec.

The foam cover that had protected against the shock of impact was ejected and four petals opened. Then 90sec after landing, the capsule began to transmit to its parent, which

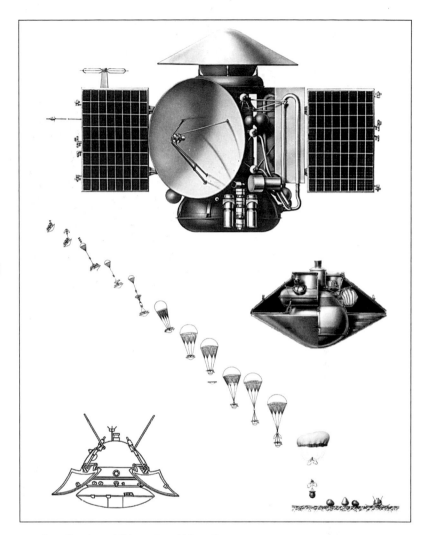

**ABOVE** The entry, descent and landing sequence for Mars 3. *(Academy of Sciences of the USSR)*

was heading for orbit insertion. When the orbiter replayed to Earth several hours later, the transmission from its lander proved to have lasted less than 20sec. The partial image is essentially noise. Furthermore, because the data collected by the lander during its descent was stored for later transmission, that was lost as well.

The consensus is that the probe failed for some reason, but some engineers believe their lander worked properly and it was the orbiter's relay that failed. The lander plus its parachute, heat shield and retrorocket might have been spotted in a super-high-resolution image obtained by a NASA orbiter in November 2007 but the identification is not certain.

## Poking out of the dust

As Mariner 9 initiated its far-encounter imaging on 10 November, one week out, the only features comprised the south polar cap and four fuzzy dark spots. One corresponded to Nix

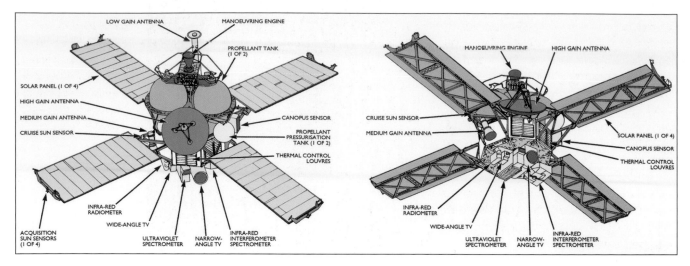

ABOVE **Details of Mariner 9.** *(NASA/JPL-Caltech/Woods)*

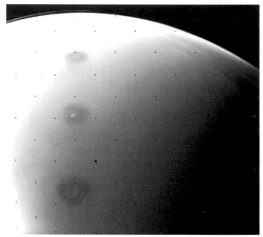

**RIGHT** A global dust storm was underway when Mariner 9 settled into orbit around Mars. The top image shows a line of three dark features that were neutrally named North Spot, Middle Spot, and South Spot. For them to be visible at all, they had to be highly elevated. As the dust began to clear, in each case a crater was revealed. These features could only be vast volcanoes with summit calderas. A fourth spot to the northwest was Nix Olympica, which proved not to be an impact crater but the largest volcano on the planet. *(NASA/JPL-Caltech)*

**BELOW** Details of the orbital activities planned for Mariner 9. *(NASA/JPL-Caltech/Woods)*

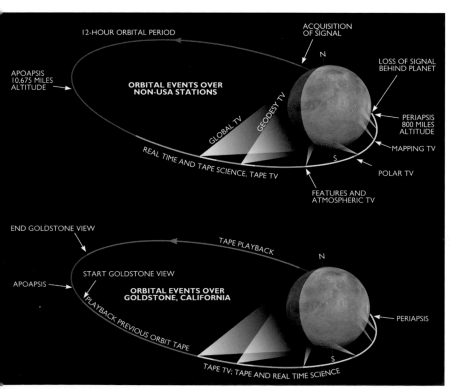

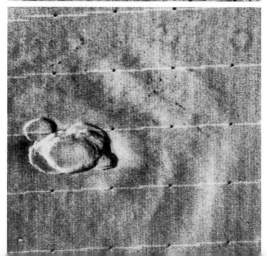

Olympica. The others, in the Tharsis area to the southeast, were spaced about 700km apart on a line running southwest to northeast which crossed the equator. They were labelled neutrally as 'North Spot', 'Middle Spot' and 'South Spot'.

On 14 November Mariner 9 became the first human artefact to achieve orbit around Mars. Each time the craft crossed the limb, its radio signal was carefully monitored to characterise the dust-laden atmosphere.

With nothing else to look at, the cameras were trained on the dark spots. As the dust started to clear, a large circular structure emerged in each case.

Once the dust had cleared, Nix Olympica was seen to be a shield volcano rising 25km above its surroundings, with its periphery marked by a scarp 8km tall. The other spots also proved to be volcanoes with summit calderas.

In contrast to the impression gained from the cratered terrain imaged by the flyby missions, it was now evident the planet had undergone intense volcanism.

Although the infrared radiometer saw no 'hot spots' in the calderas to indicate current activity, subsequent high-resolution views of lava flows on the flanks were sufficiently lacking in impacts to suggest that the vents were active in geologically 'recent' times. If so, then although the atmosphere is tenuous and dry today, from time to time it might become enriched by volcanic gases, including water vapour.

## Mars revealed, finally

In January 1972, with the atmosphere clear again, Mariner 9 refined its orbit to start mapping. This began in the southern hemisphere and migrated northward, producing a series of startling discoveries.

In retrospect, it was evident that by happenstance of timing, the three flyby missions had passed over the most lunar-like portions of the surface.

There were sinuous channels in the southern cratered terrain, some with a dendritic pattern. As the coverage crossed the equator, a canyon network was revealed that extended one-fifth of the way around the equator. Vast channels drained into the Chryse region, which was part of a low-lying plain that formed much of the northern hemisphere. The evidence of flowing

LEFT Performing a final check of the Mariner 9 spacecraft prior to installing the aerodynamic shroud of the rocket. (NASA/JPL-Caltech/KSC)

water indicated that the planet must once have possessed a hydrological cycle.

Although the discovery of geological diversity on Mars was important, the real insight from this mission was that at some point in its history the planet appeared to have undergone a change of climate. Life may have developed in the past and adapted to the altered conditions.

As Carl Sagan pointed out, "The only way to settle that argument is to land on the surface and look."

BELOW This superimposition of the Mariner 4 imaging track on a portion of a later shaded relief map shows how our new impression of the planet was biased by that mission viewing only cratered terrains. (Harland using NASA data)

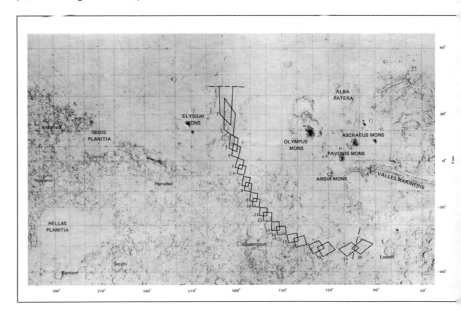

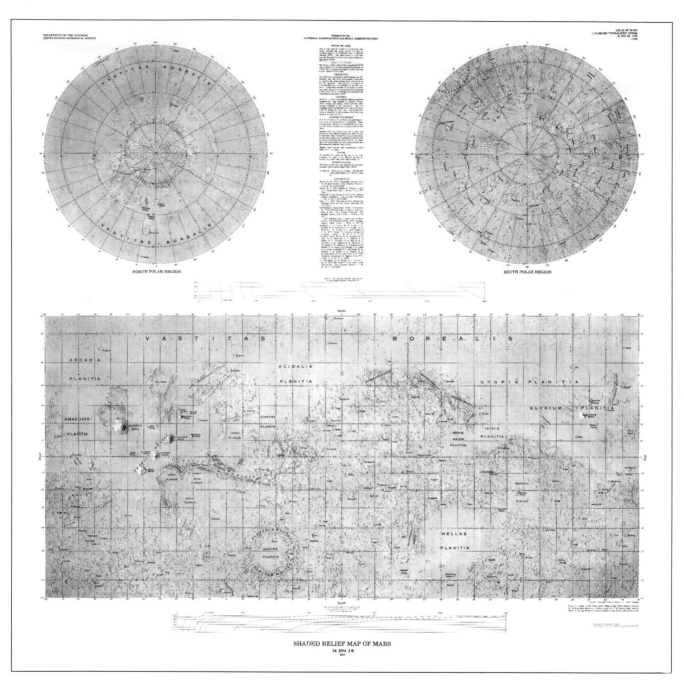

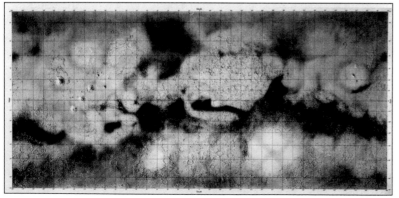

**ABOVE** After Mariner 9, the USGS produced this shaded relief map of Mars, catalogue number I-940. *(NASA/USGS)*

**LEFT** An overlay of the classical albedo features on the shaded relief map shows there to be no significant correlation with the character of the surface. *(NASA/USGS)*

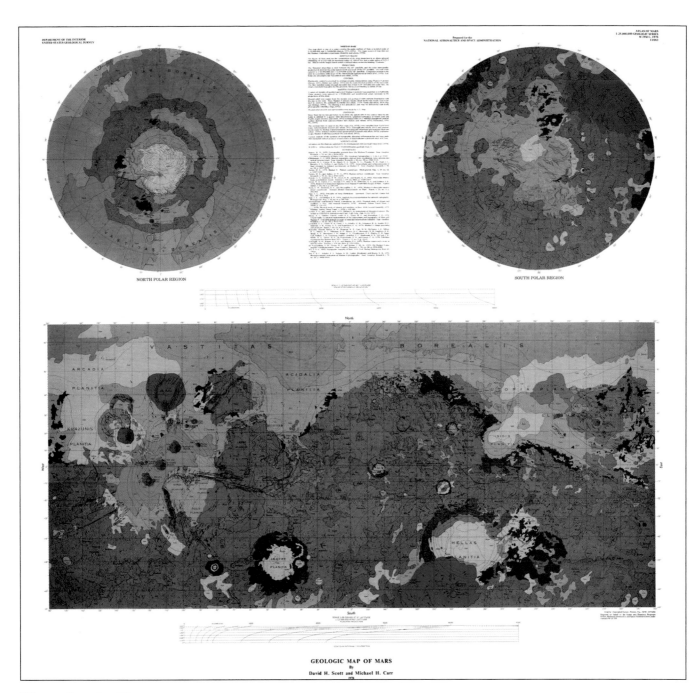

## More frustrations

The Soviets exploited the 1973 window for Mars but the planetary alignment was not as favourable as in 1971 and the landers had to be sent independently of the orbiters. The vehicles that ferried the landers were to make flybys after releasing their payloads. Unfortunately, this flotilla produced meagre results.

A propellant leak prevented Mars 4 from braking, but it provided some useful pictures while making a 2,000km flyby. Mars 5 successfully entered an orbit that was almost synchronised with the planet's rate of rotation but an air leak from its instrument compartment curtailed its operation shortly after it had begun imaging activities.

The landers fared no better. Mars 6 transmitted engineering and atmospheric data during its descent, then struck the surface at an unsurvivable 60m/sec. And Mars 7, malfunctioning immediately after being released, missed the planet.

The dejected Soviet engineers decided to give up on Mars for the immediate future.

**ABOVE** The USGS also produced this geological map of Mars based on the Mariner 9 imagery, catalogue number I-1083. *(NASA/USGS)*

## Chapter Four

# Seeking life

When astronomers observed Mars they often took it for granted the planet was inhabited by intelligent beings. After this was rejected, the consensus remained that the dark patches were vegetation. The surface revealed by the early space missions made this unlikely, so there was an imperative to find out whether the soil hosted microbial life.

**OPPOSITE** Carl Sagan stands with a model of the Viking lander in Death Valley, California during an episode of the PBS television series *Cosmos: A Personal Voyage.* *(NASA/Druyan-Sagan Associates, Inc.)*

# Test strategy

**W**hen NASA asked the National Academy of Sciences to assist in developing a strategy to find out whether life exists on Mars, the renowned molecular biologist Joshua Lederberg hosted a panel of experts in 1964 to investigate the issues. In March 1965 the draft report, *Biology and the Exploration of Mars*, said it would be reasonable to assume that life originated independently on the planet. However, it pointed out that, whereas if there were plants there would certainly be microbes, it was possible there were only microbes; hence any test for life should be directed at microbial life. The report stated, "We have reconciled ourselves to the fact that early missions should assume an Earth-like carbon-water type of biochemistry as the most likely basis of any Martian life."

In view of how biological cells function, one strategy was to seek evidence of cellular reproduction but this was a discontinuous process, the rate of which was variable from one species to the next and even in different conditions for a given species. That would make employing it as a test very difficult in the context of an exotic environment. As an ongoing process that could be measured in a number of ways, for example by changes in acidity or the evolution of gases, metabolism was more readily testable and was more

likely to produce a definitive result. The report urged a multifaceted test because, "no single criterion is fully satisfactory, especially in the interpretation of negative results". In terms of flight operations it called for "a substantial" effort to exploit the favourable launch windows in 1969, 1971, and 1973, with the first landing "in 1971, if possible" and certainly "no later than 1973".

The results from Mariner 4's flyby of Mars, a few months after the draft report was circulated, significantly diminished the prospect of there being life but didn't invalidate the strategy for seeking it.

In 1967 NASA established the Lunar and Planetary Mission Board to advise on the scientific objectives of missions, and in October 1968 this recommended making the Mars landing in 1973 with site selection based upon the mapping by orbiters in 1971. In 1970 budget constraints obliged a postponement of the new mission, named Viking, to the less favourable 1975 launch window.

## The Viking landers

**T**he mission architecture called for the lander to be ferried into orbit around Mars and then released to make the landing. The orbiter was very similar to Mariner 9 but had larger solar panels and a rocket engine for trajectory corrections during the interplanetary cruise,

**RIGHT The eccentricity of the orbit of Mars causes the distance between it and Earth when at opposition to vary considerably. Mission planning is influenced by the energy requirements of interplanetary travel, the situation being most favourable when the opposition is closest. As is apparent from this plot spanning 1965 to 1978, the situation in 1971, when Mariner 9 became the first spacecraft to orbit the planet, was particularly favourable. The original hope was to launch the Vikings in 1973, but that proved impracticable. The launch window in 1975 was rather less favourable but a bigger rocket made up the energy requirements.** *(NASA/Woods)*

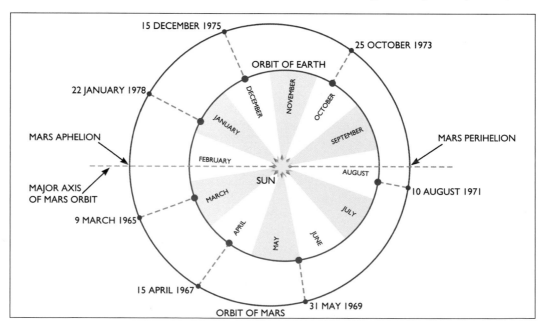

the orbital insertion burn, and such orbital manoeuvres as would be required to maintain the desired ground track timing.

The scan platform had three bore-sighted instruments. The imaging system had two television cameras set side-by-side. Each had a Cassegrain telescope with a 475mm focal length for a field of view of 1.54° x 1.69° that would cover a surface area of 40 x 44km at the planned 1,500km periapsis. The vidicon image consisted of 1,056 lines of 1,182 pixels with a 7-bit greyscale. As the acquisition of a frame and its transfer to magnetic tape would take 8.96sec, alternating the two cameras would allow a frame to be exposed every 4.48sec. In combination with the spacecraft's travel along its orbit, this alternation would record a swathe running along the ground track.

There was also a spectrometer to measure the diffuse reflection of sunlight at a wavelength of 1.38μm to determine how much water was present as vapour in the atmosphere. At periapsis this saw a footprint of 3 x 20km which was split into 15 rectangles by a nodding mirror, providing a full set of readings every 4.48sec; the same as the imaging rate. The spectrometer had a sensitivity of better than 1 precipitable micron, which is how the vapour content of an atmosphere is scaled. Because most of the water vapour on Mars is in the lowest kilometre, it would be possible to interpret the readings in terms of terrain elevation.

To the right of the scan platform was the infrared thermal mapper. During a 57sec interval it would measure a series of spots, each 8km across at periapsis, arranged in a chevron, then a mirror would flip the field of view to 'cold' space to set a zero point. The entire cycle took 1.25min. As large rocks retain heat for longer than fine-grained material, diurnal temperature gradients would offer insights into whether the surface was sandy or rocky. This information would assist in investigating suitable targets for the lander.

Once released, the lander was to make an autonomous descent, land, and operate on the surface for a minimum of 90 sols. If all went well, the computer would be updated in 3-sol cycles with a list of activities to execute. If the lander proved unable to receive commands from Earth, it was to execute a preloaded

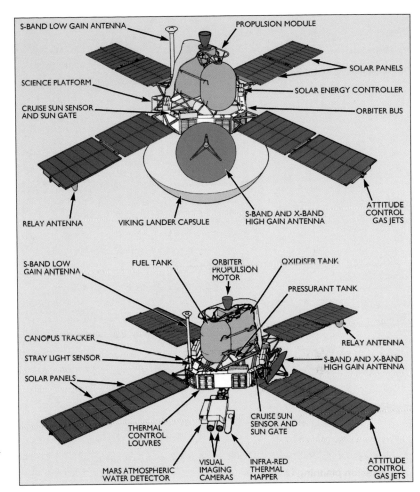

**ABOVE** Details of the Viking orbiter/lander. *(NASA/JPL-Caltech/Woods)*

**BELOW** A Titan-III-Centaur lifts off on 20 August 1975 to send the Viking 1 spacecraft to Mars. *(NASA/KSC)*

sequence of 60 sols. Although it would be possible to transmit directly to Earth, this would be at the slow data rate of 1kbit/sec. The normal mode of operating would be to transmit to the orbiter at 16kbit/sec, but this would be possible only during a window lasting at most 32min as the orbiter made a pass over the site. The orbiter would either relay the uplink to Earth in real time or store it on tape.

At times when the lander was unable to transmit to either the orbiter or Earth, it would store its data on tape for later replay.

A treaty signed in 1967 required that spacecraft sent to Mars be sterilised to preclude contaminating the planet with terrestrial microbes and creating a 'false positive' for the life-detecting instruments. It was decided to seal the lander in a bioshield and bake it for 24hr at 113°C, which tests suggested would reduce the likelihood of a single microbe hitching a ride to the Martian surface to no greater than 1 in 10,000. The challenge for the engineers was to design the systems to withstand this harsh treatment.

## Targeting

Since water vapour would be more abundant in the summer hemisphere, with its retreating polar cap, the targets for both of the Viking landers were in the northern hemisphere.

The operational constraint was that whilst the longitude of periapsis could be adjusted, there would be insufficient propellant to alter the latitude once this was established. For Viking 1, both the prime and backup sites were at 22° north. The prime site was on the sedimentary plain of Chryse Planitia at 35° of longitude and the backup was in the Elysium volcanic province at 255°.

Selecting a landing site was not simply a matter of choosing a point, because the intrinsic uncertainties of the atmospheric entry process meant the target was an elliptical 'footprint' that extended 120km in the direction of travel and 25km to either side of that track. If the lander was aimed at the centre of the ellipse there was a 99% likelihood of its reaching the surface within this perimeter.

As Mariner 9 had generally been able to resolve surface features no smaller than about 1km wide, the Viking targets had been

surveyed using a large radio telescope serving as a radar.

Although a wavelength of 13cm offered insight into the nature of the terrain on the 1m scale, the fact that the reflection was averaged over a wide area meant that a strong signal did not necessarily imply a total absence of boulders. Since the belly of the lander had only 22cm of clearance, it would still be wrecked if it were to come down on the only boulder on an otherwise smooth plain. The Viking orbiters were therefore to 'certify' the candidate sites to enable a selection to be made before releasing the landers.

Viking 1 entered orbit around Mars on 19 June 1976. It took its first picture of Chryse on 22 June, and the 200m resolution revealed there to be a profusion of small craters. This was bad news, because impacts throw out blankets of rocky ejecta. It had been planned to land on 4 July to mark the American Bicentennial but after additional imagery showed craters, channels, and cliff-edged mesas all across the prime site, the space agency announced on 27 June that the historic landing attempt would be postponed.

The pressure on the site selectors was intense because the landing had to be attempted before 22 July, when Viking 2, still inbound, would have to commit to a latitude for the periapsis of its initial orbit.

A manoeuvre enabled Viking 1 to investigate a site 250km northwest of the nominal target, since the terrain appeared to become smoother in that direction, but this was still too rough. Attention then switched to a site 580km farther west, where there were fewer fresh-looking craters, and on 13 July NASA decided to attempt to land at 22.5° north and 47.4° of longitude on 20 July.

## Descent

When released by explosive bolts, the lander operated autonomously. The time taken for a radio signal to travel from Mars to Earth was 18min. If anything went awry its masters were powerless to intervene; they were merely spectators. The separation was at an altitude of 5,000km in an elliptical orbit. The lander made a de-orbit burn of 156m/sec, to enable it to contact the atmosphere on a trajectory

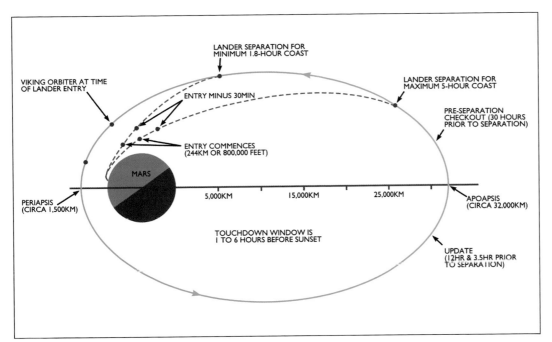

VIKING ORBITER AT TIME OF LANDER ENTRY

LANDER SEPARATION FOR MINIMUM 1.8-HOUR COAST

ENTRY MINUS 30MIN

LANDER SEPARATION FOR MAXIMUM 5-HOUR COAST

PRE-SEPARATION CHECKOUT (30 HOURS PRIOR TO SEPARATION)

ENTRY COMMENCES (244KM OR 800,000 FEET)

MARS

PERIAPSIS (CIRCA 1,500KM)

5,000KM     15,000KM     25,000KM

APOAPSIS (CIRCA 32,000KM)

TOUCHDOWN WINDOW IS 1 TO 6 HOURS BEFORE SUNSET

UPDATE (12HR & 3.5HR PRIOR TO SEPARATION)

**LEFT** Once in orbit around Mars the Viking spacecraft would release its lander on a trajectory that would cause it to penetrate the planet's atmosphere. *(NASA/JPL-Caltech/Woods)*

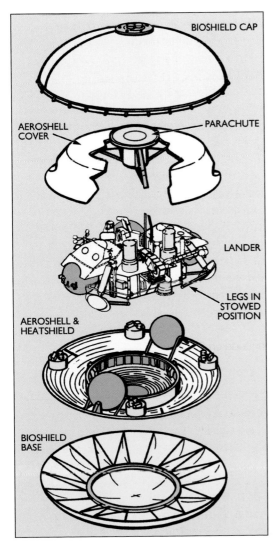

BIOSHIELD CAP

AEROSHELL COVER

PARACHUTE

LANDER

LEGS IN STOWED POSITION

AEROSHELL & HEATSHIELD

BIOSHIELD BASE

depressed 16° below horizontal at a point which, if things went well, would result in a landing within the target ellipse.

As the lander penetrated the atmosphere, a suite of sensors on its aeroshell started to measure the chemical composition, temperature, and pressure.

The public auditorium at JPL was packed. In addition to 400 journalists from around the world, 1,800 invited guests were watching a closed-circuit television view showing the

**FAR LEFT** The Viking atmospheric entry system. *(NASA/JPL-Caltech/Woods)*

**BELOW** Preparing a Viking lander. *(NASA/KSC)*

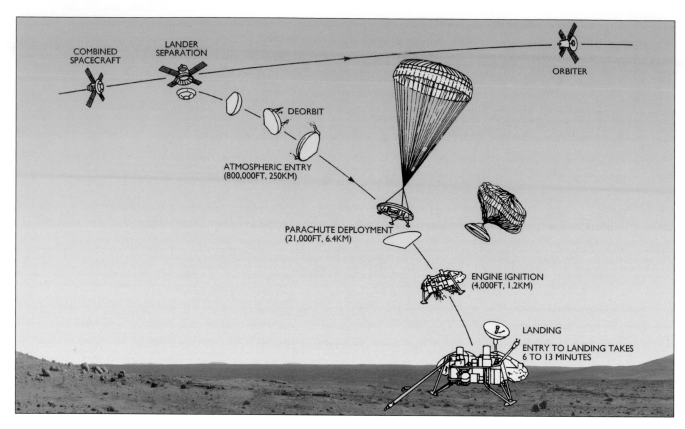

**ABOVE** The Viking entry, descent and landing sequence. *(NASA/JPL-Caltech/Woods)*

control room, where Albert Hibbs, one of the mission planners, was providing the commentary.

"We're coming down," Hibbs reported, as the telemetry downlink showed the vehicle descending though an altitude of 24km, having survived a peak dynamic load of 8.4g. "It's a long period of almost flat glide to get rid of some more of the speed before the parachute comes out."

As the vehicle slowed to 1km/sec, the load diminished to 0.8g. At an altitude of 6km a mortar in the rear section of the aeroshell deployed a parachute with a diameter of 16m and the forward part of the shell was jettisoned. A few seconds later the three legs of the lander were deployed. At an altitude of 1,400m, having slowed the rate of descent to 54m/sec, the parachute was released and a trio of rocket engines ignited to execute the terminal phase.

An inertial reference and a radar altimeter enabled the vehicle's computer to control the rate of descent by varying the thrust. A requirement for the design of the engines was to minimise their disturbance of the surface material, because the biologists did not want the exhaust gases to contaminate the soil that

was to be tested for life. To disperse the plume, each engine had 18 small nozzles that pointed in a range of downward directions. The rockets were to slow the rate of descent to 2m/sec for contact. As the system would be shut down by sensors in the base of the three footpads, the first leg to make contact would terminate the controlled phase of the descent. Thereafter the vehicle would fall under gravity. Upon receiving the contact signal the computer would also switch the telemetry from 4 to 16kbit/sec. Seeing this transition, Hibbs joyfully announced: "We have touchdown!"

This was an historic moment because, as Carl Sagan had wryly pointed out, "You can only land on Mars for the *first time* once."

## On Mars at Chryse

It was late afternoon on Chryse Planitia. In fact the landing site was beyond the limb as seen from Earth, and the lander's signal was being relayed by the orbiter as it passed overhead and continued on towards the horizon. The lander's first assignment was to take pictures of the site as a contingency against something untoward occurring during the Martian night.

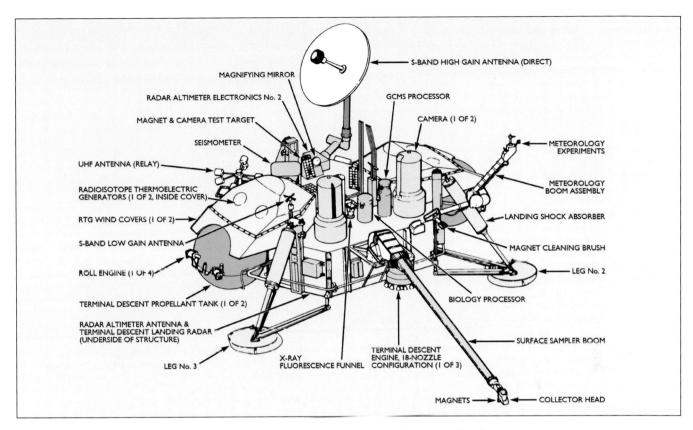

MAGNIFYING MIRROR

RADAR ALTIMETER ELECTRONICS No. 2

MAGNET & CAMERA TEST TARGET

SEISMOMETER

UHF ANTENNA (RELAY)

RADIOISOTOPE THERMOELECTRIC GENERATORS (I OF 2, INSIDE COVER)

RTG WIND COVERS (I OF 2)

S-BAND LOW GAIN ANTENNA

ROLL ENGINE (I OF 4)

TERMINAL DESCENT PROPELLANT TANK (I OF 2)

RADAR ALTIMETER ANTENNA & TERMINAL DESCENT LANDING RADAR (UNDERSIDE OF STRUCTURE)

LEG No. 3

X-RAY FLUORESCENCE FUNNEL

TERMINAL DESCENT ENGINE, 18-NOZZLE CONFIGURATION (I OF 3)

S-BAND HIGH GAIN ANTENNA (DIRECT)

GCMS PROCESSOR

CAMERA (I OF 2)

METEOROLOGY EXPERIMENTS

METEOROLOGY BOOM ASSEMBLY

LANDING SHOCK ABSORBER

MAGNET CLEANING BRUSH

LEG No. 2

BIOLOGY PROCESSOR

SURFACE SAMPLER BOOM

MAGNETS — COLLECTOR HEAD

It sent two monochome pictures to the orbiter, which taped them for relay to Earth upon emerging from its passage behind the planet.

Because of how the camera operated, the image was built up as a series of vertical lines which slowly progressed to the right on a display screen. Everyone was eager for their first view of Mars at ground level.

The first few lines showed streaks resulting from the slow-scanning camera catching the swirling dust, but once this had settled the clarity was startling. As the image grew, a small rock appeared sitting on a level surface of fine-grained material. Then additional rocks interspersed with pebbles. There were signs of aeolian erosion by dust. Some rocks featured small holes resembling vesicles. One of the 0.3m circular footpads appeared several

BELOW The first ground-level view of Mars! This was transmitted by Viking 1 minutes after landing as a series of vertical lines, starting on the left. In the initial scans, dust stirred up by the braking rockets was still in the process of settling. (NASA/JPL-Caltech)

minutes into the process. It was resting cleanly on the surface, undamaged, with some dust on its concave upper surface. It was a stunning sight!

Over the next 30min a panorama built up showing a view out to the horizon some 3km distant, the line of which showed that the lander had come to rest on level ground. The landscape was strewn with rocks, some of which were rather large, and there were sand dunes. Geologists noted the striking similarity to the high desert of the American Southwest.

On 21 July (sol 1) the first colour panorama was taken. Although there were colour charts on the lander, the picture was released to an eager public without being properly calibrated and it showed the sky as pale blue. A colour-balanced version released the following day revealed the sky to be salmon pink. In Earth's atmosphere, Rayleigh scattering by air molecules diffuses blue sunlight, making the sky blue. The much more rarefied Martian atmosphere does not scatter blue very efficiently, but (as occurs on Earth at sunset when the air is dusty) the dust motes scatter the red end of the spectrum.

## VIKING LANDER IMAGING SYSTEM

**BELOW Details of the line-scan imaging system used by the Viking lander.** *(NASA/JPL-Caltech/Woods)*

The imaging system had two cameras mounted on top of the lander, spaced 1m apart for stereoscopic viewing. As the lander was initially intended to have been capable of transmitting its imagery directly in real time, the scan rate of the novel cameras was matched to the transmission rate. When a tape recorder was later added, it was too late to speed up the imaging process.

Each camera had an upward-aimed photodiode at the base of a cylinder. The optics were above the photodiodes and topped by a mirror that rotated to provide a vertical sweep of 512 pixels, each a 6-bit word. The cylinder rotated axially after each sweep to extend the image horizontally, line by line. As the cameras could rotate almost 360° they could view across the top of the deck as well as outward and, between them, take panoramic views of the landing site. However, doing so took a long time. In fact, when a test version of the camera was set up to take a panoramic picture of a large group of project staff, those who were in shot at any given time had to stand very still and it took so long to complete the scan that by the time the line of sight reached the people at the far end of the group, those at the start had long-since departed!

There were a dozen diodes: one for monochrome panoramas; a red, green, and blue for subsequent composition into colour; three infrared; four at different focal points for high-resolution monochrome imagery; and one of low sensitivity for when the Sun was in the field of view.

Although this type of camera may seem rather primitive in comparison to a modern CCD imager, at that time it was state of the art.

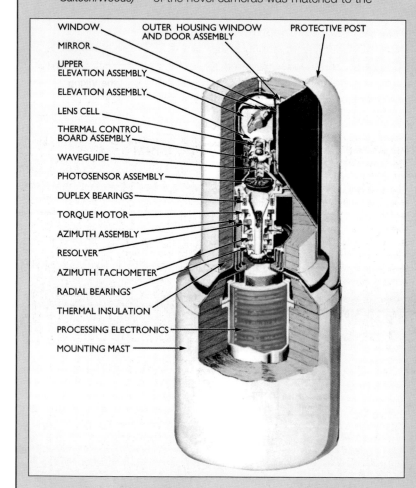

WINDOW
MIRROR
UPPER ELEVATION ASSEMBLY
ELEVATION ASSEMBLY
LENS CELL
THERMAL CONTROL BOARD ASSEMBLY
WAVEGUIDE
PHOTOSENSOR ASSEMBLY
DUPLEX BEARINGS
TORQUE MOTOR
AZIMUTH ASSEMBLY
RESOLVER
AZIMUTH TACHOMETER
RADIAL BEARINGS
THERMAL INSULATION
PROCESSING ELECTRONICS
MOUNTING MAST

OUTER HOUSING WINDOW AND DOOR ASSEMBLY
PROTECTIVE POST

The lander was equipped with a robotic arm with which to retrieve soil for the science experiments. It had a reach of 3m and could swivel through a horizontal arc of 230°, elevate 35° and dip 50°. At its end was the collector head. To gather a sample the lid would be raised, the boom driven into the topsoil, the lid closed, the boom retracted, the head rotated to dump the sample onto the lid, and then vibrated to encourage finely grained material to fall through small holes into the appropriate sample inlet.

The arm delivered some soil to the X-ray fluorescence spectrometer. It was irradiated to cause the nuclei of the constituent atoms of that material to emit at characteristic frequencies in order to generate a series of electrical pulses in the detectors in proportion to the energy of the irradiation. The pulses were counted over a succession of intervals. As X-ray analysis is a statistical process it was a case of increasing the ratio of signal to noise for the data. This experiment gave the composition of the material in terms of the abundances of elements. On this basis the team inferred the likely mineralogical abundances. The main elements present were iron, calcium, silicon, titanium and aluminium. One likely analogue was Hawaiian basalt. As the rock was gradually weathered to finely grained particles, this would create the dust that is so readily blown about.

**LEFT** To provide colour imagery, the lander transmitted several images of the same scene using filters for different wavelengths. In combining the images the JPL engineers presumed that the sky would be blue and released the image on the left, but after checking the calibration they realised the sky was 'salmon pink'.
*(NASA/JPL-Caltech)*

**BELOW** A picture of rocks near the Viking 1 lander.
*(NASA/JPL-Caltech)*

**BOTTOM** Part of a horizon panorama at the Viking 1 landing site.
*(NASA/JPL-Caltech)*

At only 15.5kg, the Viking biology package was a miracle of miniaturisation. It incorporated four experiments to seek evidence of microbial life in the Martian soil. The samples would be excavated and delivered by the robotic arm of the lander. The lander was powered by a radioisotope thermal generator and thus could run the experiments independently of the solar cycle.

One instrument was to separate vapour components chemically by using gas chromatography then feed its output to a mass spectrometer that would measure the molecular weight of each chemical. It could detect molecules at several parts per billion. There was particular interest in identifying organic chemicals that were essential for biology.

The gas exchange experiment was to look for gases issued by an incubated soil sample by first replacing the Martian atmosphere with the inert gas helium. It would then apply a fluid of organic and inorganic nutrients and supplements to a sample, initially with just nutrients added and then with water too. The instrument would periodically sample the atmosphere of the incubation chamber and feed it to the gas chromatograph to measure the concentrations of gases, one or other of which were expected to be consumed or released by metabolising organisms.

The labelled release experiment held the most promise for the exobiologists. A sample of soil was to be inoculated with a drop of

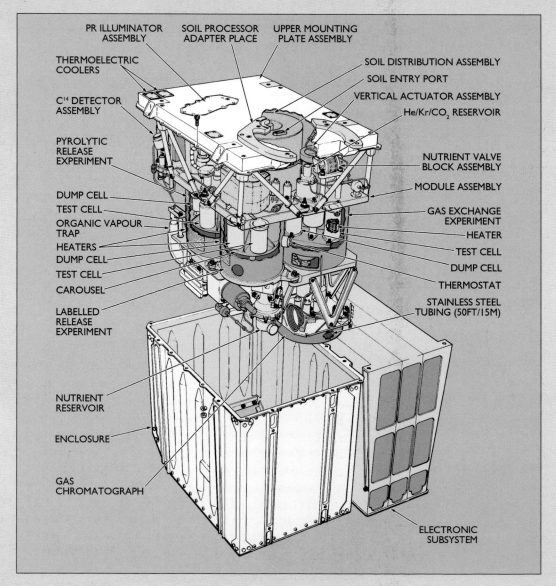

**RIGHT A schematic of the biology package developed for the Viking lander.** *(NASA/JPL-Caltech/TRW Systems/Woods)*

PR ILLUMINATOR ASSEMBLY

SOIL PROCESSOR ADAPTER PLACE

UPPER MOUNTING PLATE ASSEMBLY

THERMOELECTRIC COOLERS

$C^{14}$ DETECTOR ASSEMBLY

PYROLYTIC RELEASE EXPERIMENT

DUMP CELL

TEST CELL

ORGANIC VAPOUR TRAP

HEATERS

DUMP CELL

TEST CELL

CAROUSEL

LABELLED RELEASE EXPERIMENT

NUTRIENT RESERVOIR

ENCLOSURE

GAS CHROMATOGRAPH

SOIL DISTRIBUTION ASSEMBLY

SOIL ENTRY PORT

VERTICAL ACTUATOR ASSEMBLY

He/Kr/$CO_2$ RESERVOIR

NUTRIENT VALVE BLOCK ASSEMBLY

MODULE ASSEMBLY

GAS EXCHANGE EXPERIMENT

HEATER

TEST CELL

DUMP CELL

THERMOSTAT

STAINLESS STEEL TUBING (50FT/15M)

ELECTRONIC SUBSYSTEM

very dilute aqueous nutrient solution tagged with radioactive carbon-14. The air above the soil was then to be monitored for the evolution of appropriately radioactive carbon dioxide gas. That would be evidence that micro-organisms in the soil had metabolised one or more of the nutrients.

The pyrolytic release experiment would use light, water, and an atmosphere of carbon monoxide and carbon dioxide, using carbon-14 in the carbon-bearing gases. If there were photosynthetic organisms present, it was believed that they would take in some of the carbon as biomass by the process of carbon 'fixation'. After several days of incubation, the experiment would remove the gases, bake the remaining soil at 650°C and measure the radioactivity of the products. If any of the radioactive carbon had been converted to biomass, it would be vaporised during heating and its detection would be evidence for metabolism. In the event of achieving a positive response, a duplicate sample of the same soil was to be heated to sterilise it prior to testing. If this were to produce a similar outcome, it would imply that the activity was due to inorganic chemical reactions rather than biology.

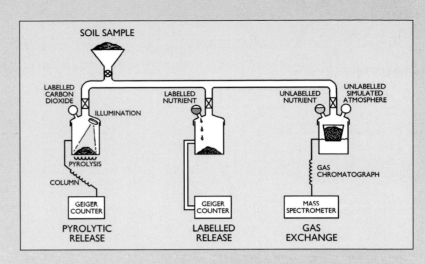

**ABOVE** The three Viking biology experiments. *(NASA/Woods)*

**BELOW** The Gas Exchange experiment. *(NASA/Woods)*

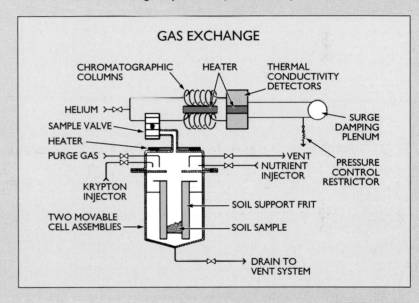

**BELOW** The Labelled Release experiment. *(NASA/Woods)*

**BELOW RIGHT** The Pyrolytic Release experiment. *(NASA/Woods)*

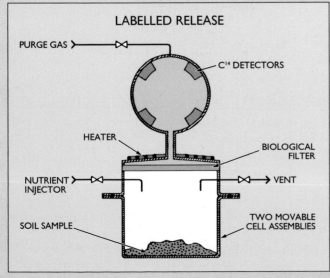

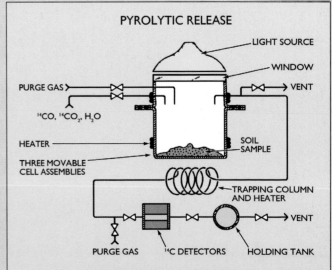

**ABOVE Another part of the Viking 1 panorama.** (NASA/JPL-Caltech)

**BELOW By repeatedly scanning a single line of sight, Viking 1 recorded how the illumination changed over time. Note the dust storm on sol 1742.** (NASA/JPL-Caltech/ Oliver de Goursac)

The lander carried a sophisticated suite of instruments to test whether there was life in the Martian soil.

The gas exchange experiment was initiated on 29 July. This assumed that Martian life would resemble terrestrial life, and if the planet underwent episodic climate variations then its microbes might go dormant during the long, cold, arid times, awaiting a resumption of benign conditions, which the experiment was to provide. The task was to determine whether biological metabolism modified the composition of the gases in the test chamber. The sample was to be offered an aqueous solution of nutrient named 'chicken soup' that had almost everything a terrestrial microbe might consume – amino acids, purines, pyrimidines, organic acids, vitamins, and minerals. Since water vapour was not stable at the surface,

the pressure in the chamber had to be significantly greater than local to prevent the nutrient from breaking down. The experiment involved two stages.

First, a mix of carbon dioxide and krypton was added to the chamber and a mist of nutrient was supplied to expose the sample to water vapour. After 2hr a sample of the gases present in the chamber was sent to a gas chromatograph and mass spectrometer. This would define the benchmark for comparison with analyses at various stages of the incubation seeking metabolic products such as hydrogen, oxygen, nitrogen, carbon dioxide, and methane. This first measurement gave a surprisingly large peak for oxygen.

Was this an orgy of metabolism as the nutrient awakened dormant microbes? The scientists couldn't proclaim life as soon as they

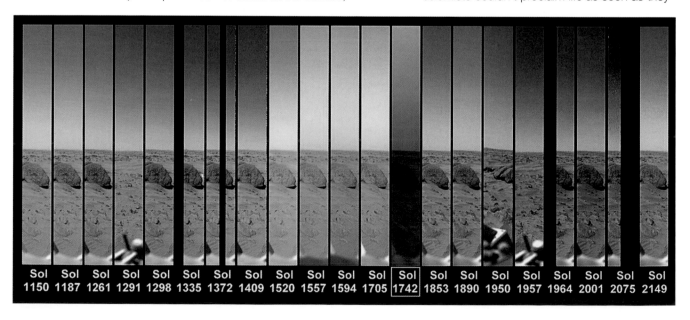

Sol 1150 | Sol 1187 | Sol 1261 | Sol 1291 | Sol 1298 | Sol 1335 | Sol 1372 | Sol 1409 | Sol 1520 | Sol 1557 | Sol 1594 | Sol 1705 | Sol 1742 | Sol 1853 | Sol 1890 | Sol 1950 | Sol 1957 | Sol 1964 | Sol 2001 | Sol 2075 | Sol 2149

saw a response which could imply biological activity; they had to wait for a reaction that *could only* be explained by biology.

The 'humid mode' of the experiment was to last a week. For this, the sample was suspended in a porous cup above the nutrient, without ever being in direct contact with the solution. The release of oxygen slowed considerably. The rapid release of oxygen followed by a decline hinted at an intense but brief inorganic reaction between the soil and the water in the nutrient. Free hydroxyl ions from ultraviolet dissociation of water vapour at low altitude would build up 'peroxy' compounds in the chilly, dry soil. Peroxides, superoxides, and ozonides are all strong oxidising agents which would rapidly break down into water and gaseous oxygen in the presence of a significant amount of water vapour. Did this explain why so much oxygen was evolved by the sample when presented with a mist of water vapour?

On 5 August the experiment advanced to its 'wet mode' with the injection of nutrient directly into the sample. This caused a third of the carbon dioxide in the chamber to be dissolved by either the sample or the water, then as the six-month incubation proceeded, it gradually returned to its initial level and activity ceased. The observed uptake of carbon dioxide was explicable as water causing peroxy compounds in the soil to draw in carbon dioxide to create either metal oxides or hydroxides, and the later slow release of this gas could have been iron oxides in the soil reacting with nutrients dissolved in the water and liberating it. Meanwhile oxygen was reclaimed rather than released, with the amount of oxygen

taken up matching some of the constituents of the nutrient. Evidently the Martian soil was extremely reactive.

The labelled release experiment was initiated on 30 July. This made fewer assumptions concerning the biochemistry of Martian life. On the assumption that life would be adapted to its environment, the nutrient was a weak 'broth' made of glycine and alanine amino acids, and formic, glycolic and lactic acids in the form of salts in distilled water. The premise of the test was that if microbes consumed the nutrient, this metabolic activity would produce gases such as carbon dioxide or methane that could be detected by labelled carbon-14 using a Geiger counter. In contrast to the gas exchange experiment, this apparatus wasn't to identify the particular gases produced. A mist of nutrient was introduced into the chamber to moisten the soil and helium was injected to maintain the pressure to prevent the nutrient breaking down.

A rapid rise to 10,000 counts per minute indicated that a large amount of gas was evolved as soon as the nutrient was added, but by 2 August it was apparent that there was no exponential increase to indicate growth. This behaviour did not match either the predicted biological or chemical responses. After the counts had levelled off, indicating that the release of radioactive gases had ceased, a second injection of nutrient was undertaken. If the initial evolution of gas had been due to microbial metabolism of the nutrient, there ought to have been a second rise, but the rate rapidly fell to 8,000 counts and levelled off, implying either the microbes had expired after their initial feast or the reaction was inorganic chemistry.

Given the evidence for peroxy compounds in the sample, hydrogen peroxide could have oxidised constituents of the nutrient to carbon dioxide and water. The amount of radioactive carbon dioxide was only slightly less than would have been liberated if all of the formic acid had been consumed in this manner. If the oxygen emitted in the gas exchange experiment was due to the dissociation of peroxides in the soil by the water vapour of the nutrient mist, then the water produced in the experiment by the breakdown of formic acid ought to have decomposed all of the peroxy compounds at the first injection of nutrient, with no further radioactive gas being issued in response to the second injection. The fact that the count actually fell with the second injection implied some of the carbon-14 was being reclaimed, probably when carbon dioxide was absorbed by the water in the nutrient.

The pyrolytic release experiment was initiated on 28 July. This assumed that any Martian microbes would be adapted to the environment but also need to 'fix' carbon from the atmosphere. The experiment was to seek evidence of synthesis of organic matter in conditions as close as possible to the local environment. The soil was sealed into the test chamber and the Martian air drawn out and replaced by a representative mixture of carbon monoxide and carbon dioxide labelled with carbon-14. During incubation, the chamber was illuminated by a xenon arc which simulated sunlight at the Martian surface, minus the ultraviolet component.

After 120hr, the lamp was to be switched off and the gases flushed out to be analysed by a Geiger counter to produce the 'first peak' for the carbon-14 still in the air. To determine whether microbes had assimilated carbon, the sample was to be heated to 640°C in order to pyrolyse the organic molecules and release the carbon-14 as carbon dioxide, which would be measured as the 'second peak'. A high second peak would support a biological interpretation, but a low one would indicate few, if any, microbes (with the strength of this conclusion depending on the value of the peak). On 7 August it was announced that the second peak was strong, but again there was ambiguity. The results argued against the peroxides that were favoured as a chemical interpretation of the gas exchange's response, since the pyrolytic release had produced a reducing as opposed to an oxidising reaction, and peroxides would have tended to decrease the scale of the second peak.

Next the labelled release and pyrolytic release experiments were repeated as 'control' tests in which, prior to incubation, the samples were heated sufficiently to kill microbes but not to inhibit most inorganic chemical reactions.

On 20 August it was announced that although the control test of the labelled release could have ruled out biology, it didn't do so. In

fact, the radioactivity had rapidly risen to 2,200 counts, fallen sharply and then levelled off at 1,200 counts.

Gilbert V. Levin, who devised the experiment, was encouraged, "If we'd run this experiment in the parking lot at JPL [and seen these two curves] we'd have concluded that life is present in the sample." But if it was a chemical reaction it was one that was disabled by heat. "We've significantly narrowed the range of possible chemical explanations."

If the initial carbon assimilation of the pyrolytic release had been caused by biology then the sterile sample ought to have been negative. In reality there was assimilation, albeit at a much reduced level. This argued for a chemical reaction involving a number of reactants, only some of which had been inhibited by prior heating.

When the gas exchange experiment had finished its six-month incubation the container was reloaded with fresh soil that was sterilised before rehumidification. The fact that half of the initial amount of oxygen was released indicated that there was a non-biological response. Because the pre-heating would have dissociated hydrogen peroxide, this indicated the reaction was the result of a more thermally stable superoxide.

The initial analysis by the gas chromatograph and mass spectrometer was on 6 August. The sample was first heated to 200°C to drive most of the water out of the hydrated minerals, but there was surprisingly little water produced. Next, the sample was heated to 500°C to volatise organic molecules. The astonishing fact that there were no organics was reported on 13 August. However, it was thought the analysis had been complicated by the delayed release of water.

Another sample was tested on 21 August to gain further insight. A significant amount of water was liberated at 350°C, but despite the improved sensitivity in the second stage, the organics present, if any, were below the 10–100 parts per billion detection limit. On the other hand, there would have had to have been at least 1 million microbes in the sample for their organic material to be detected at that level of sensitivity. A typical temperate sample of terrestrial soil can contain hundreds of millions of bacteria per cubic centimetre. If only the living cells were present for analysis, then 1 million bacteria would have been far too few for the instrument to detect. In terrestrial soil, the amount of dead organic matter often outweighs the living material by a factor of 10,000. If Martian microbes were the same, they would probably have been able to be detected by the organic waste and dead cells that they left. But if such microbes recycled their waste and solar ultraviolet destroyed the dead cells there could have been microbes present to cause the reactions observed by the biology package without the spectrometer being able to detect them.

To investigate further, Levin later processed samples of soil from Antarctica and confirmed that soil which the gas chromatograph and mass spectrometer reported to be devoid of organics had nevertheless hosted sufficient bacteria to replicate the results of the labelled release experiment on Mars. Although this hypothesis explained the conflicting results, there was no proof that it was true.

## At Utopia too

Arriving on 7 August, Viking 2 entered an orbit with its periapsis at 44° north, as planned. For operational reasons the two landers had been assigned targets on opposite sides of the planet.

A reconnaissance by the Viking 1 orbiter had found the intended landing site for its partner in the Cydonia region to be too rough, so the new arrival's first task was to search for somewhere more suitable. If necessary NASA would postpone this second surface mission until after Mars had passed through the conjunction in

November, because communications would be impaired whilst the planet was on the far side of the Sun.

The inability of terrestrial radar to produce information on the smoothness of sites at the greater latitude was compensated by the infrared thermal mapper of the orbiter which, at periapsis, gave insight into whether the surface was blocky.

A site at 47.9° north and 225.9° of longitude on Utopia Planitia was selected and the Viking 2 landing was scheduled for 3 September.

The separation of the lander seemed nominal but several seconds later the orbiter started to suffer attitude control problems which prevented

ABOVE A Viking orbiter releasing its lander. *(NASA-JPL-Caltech/Don Davis)*

BELOW The first image provided by Viking 2. *(NASA/JPL-Caltech)*

the high-gain antenna from relaying the lander's telemetry. As a result, the engineers at JPL were even more anxious than for the first landing. They could only monitor the low-gain signal from the orbiter for an indication that the lander had boosted its transmission rate from 4 to 16kbit/sec on touchdown. The tension in the control room was electric when the predicted time of transition passed, but 21sec later the signal rate increased and the celebrations began.

As Viking 2 pursued its post-landing activities, the engineers concluded that the orbiter had been struck by one of the explosive bolts in the separation of the lander. Fortunately the spacecraft was able to recover in time to record the first pictures from the lander and these were relayed once the high-gain link was re-established.

Contrary to expectation, Utopia was strewn with rocks, one of which was so close to the documented footpad that it may well have been nudged aside during the landing. The chemical analysis of the soil at this site proved to be very similar to that at Chryse Planitia. The biology instruments imposed tighter constraints on putative inorganic chemistry, but did not resolve the debate.

## Disputed results

Gilbert Levin argued, "The accretion of evidence has been more compatible with biology than with chemistry – each new test result has made it more difficult to come up with a chemical explanation, but each new result has continued to allow for biology." Levin insisted that if a terrestrial sample had given the results seen on Mars, "We'd unhesitatingly have described [it] as biological."

Vance I. Oyama, who created the gas exchange experiment, was sceptical: "There was no *need* to invoke biological processes."

Norman H. Horowitz of the pyrolytic release experiment agreed, but pointed out it was "impossible to prove that any of the reactions … were *not* biological in origin".

So prior to the Viking landings no one knew whether there was life on Mars, and, sadly, no one knew afterwards either!

The biology team leader Harold P. Klein later concluded that the assumption that Martian microbes must be similar to terrestrial ones ought to be dismissed, and scientists should review whether the Viking data suggested any clues as to "whether there might be some less obvious kind of life on Mars".

In fact, even as the Vikings were seeking carbon-synthesising microbes on Mars, biologists on Earth were discovering the first examples of an entirely new class of microbial life.

## Extremophiles

Earth was accreted from the solar nebula some 4.5 billion years ago and within around 100 million years it had cooled sufficiently for a hydrosphere to form. At that time the surface of the planet was dominated by volcanism that pumped up the atmosphere with carbon dioxide, creating a 'greenhouse' that trapped solar energy.

The oldest known rocks are about 4 billion years old, but these have been altered by subsequent processing and so can reveal nothing of the likelihood of life at that time. The first convincing evidence is in well-preserved 3.5 billion-year-old rocks at Barberton in South Africa and in the Pilbara in Australia that contain various indicators of microbial ecosystems.

It had been believed that water in the hottest of geothermal springs must be sterile but in the early 1970s microbes were discovered living in water at 85°C in Yellowstone National Park.

Then in 1977 a submersible that was investigating the rift in the Galapagos Ridge discovered a hydrothermal vent on the ocean floor which was emitting a super-heated plume of water. A similar vent located in 1979 on the East Pacific Rise was so rich in dissolved minerals that a 'chimney' had formed. Such 'black smokers' supported ecosystems that hosted many species of life, some entirely new to science.

In 1982 it was found that these isolated food chains were based on single-celled organisms that derived their energy from the nutrients in the vent. Then exceptionally well-preserved mineral textures that were indistinguishable from contemporary black smokers were found in a 3.26 billion-year-old sediment in Australia.

These thermophilic microbes became the first known members of a wholly new class of life named archea (the old ones).

Biological cells are autotrophs or heterotrophs (or perhaps both) in that they either manufacture organic compounds themselves from raw materials such as carbon dioxide (autotrophs, meaning self-feeders) or they draw organic material from their

**ABOVE** Operations by the robotic arm of the Viking 2 lander. *(NASA/JPL-Caltech)*

environment and process it into whatever they require (heterotrophs). Which type came first was disputed.

The heterotrophic origin theory was proposed in the 1920s by the Russian chemist Alexander Ivanovich Oparin, and appeared reasonable. After all, why should not the earliest life have exploited the organic compounds present in its environment?

However, opinion now favours an autotrophic origin. This conclusion was reinforced by the discovery of yet another unusual microbial ecosystem.

In the mid-1980s microbes were found in the interstices between grains of rock at a depth of 1km beneath the surface of the Piceance Basin in Colorado. There were some heterotrophs present which consumed the remains of plant detritus bound up in sedimentary rock. They resembled those that lived on the surface, but had adapted to the hot anoxic environment by remaining static in their individual niches between grains of rock. Most of the microbes, however, were autotrophs that exploited the heat and hydrocarbons which were toxic to

'conventional' life. Because these 'lived off rock' they were named lithotrophs.

Autotrophs are well represented among the archea. Those that thrive in a hydrothermal environment that is lacking both oxygen and light are collectively known as either anaerobic autotrophs or chemolithotrophs. Many require only water enriched with volcanic gases and nutrients.

The fact that none of the archean autotrophs use sunlight implies that the process of photosynthesis arose later. The ability to draw energy from sunlight was a major advance because it is more efficient.

Unrestricted to hydrothermal vents, a single-celled cyanobacteria that could use chlorophyll for photosynthesis was free to colonise the planet. They initially formed thin mats on the floors of shallow seas. Layers of rock were accreted as the mats trapped particulates in suspension and minerals that precipitated from the water. Such colonies are called stromatolites. There is disputed evidence of fossil stromatolites in the Pilbara, but no doubt about the biological origin of 3.5 billion-year-old

**BELOW** Stromatolites in Shark Bay, Western Australia. These single-celled cyanobacteria colonies are 'living fossils' from the earliest of times.

fossils at Baffin Island in Canada because they can be matched, point for point, with the 'living fossils' that still survive in the highly saline Shark Bay of Western Australia.

The development of archea on Earth had profound implications for life on Mars.

Due to the intense ultraviolet in sunlight, the absence of a magnetic field to protect the surface against solar wind particles, and the presence of oxidising chemicals in the soil, the Martian surface is nowadays inhospitable to organic chemistry but life may have originated early on, when the planet was intensely volcanic. Then when surface conditions changed, the Martian counterparts of archea could have continued to live underground.

If the Viking landers had been able to drill to a depth of several metres to obtain the samples for the biology experiments, then the outcome might have been very different.

Indeed, if the mid-1960s team led by Joshua Lederberg had known of the archea when it devised the strategy for seeking life in the soil of Mars, a rather different set of tests might have been recommended.

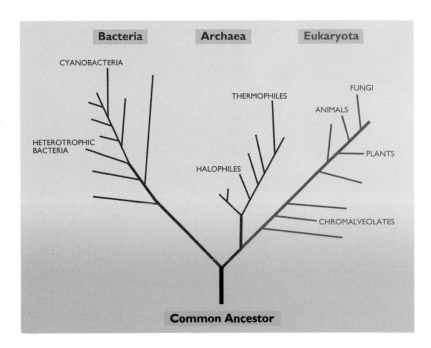

**ABOVE** The discovery of the extremophile archea introduced a third branch to our understanding of the tree of life on Earth. *(Woods)*

**BELOW** A great many colour images of Mars provided by the Viking orbiters were combined to produce a map of the planet which far surpassed that of Mariner 9. *(NASA/JPL-Caltech/USGS)*

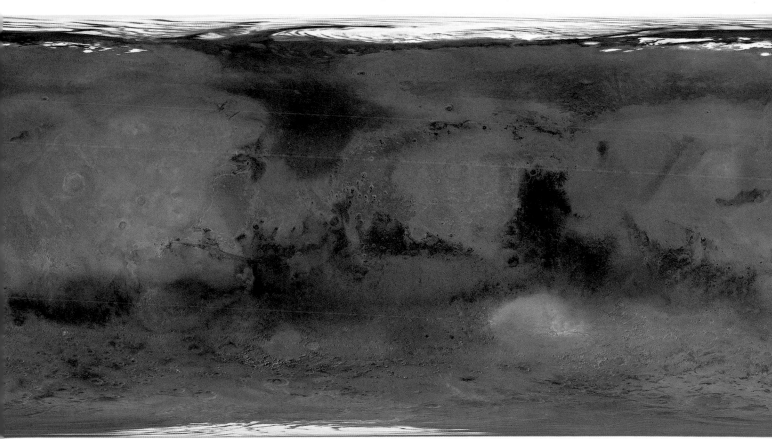

*Chapter Five*

# New views
# from orbit

Orbiters with ever more sophisticated sensors have mapped Mars in terms of the distribution of minerals and subsurface water ice, investigated the role of water in the formation of surface features, monitored seasonal variations, measured how the solar wind erodes the atmosphere, and shed light on how surface conditions changed from warm and wet to cold and arid.

**OPPOSITE An artist's depiction of Mars Reconnaissance Orbiter firing its engine to enter orbit around the planet.** *(NASA/JPL-Caltech/Corby Waste)*

# A poor start

In addition to relaying for their landers, the Viking orbiters mapped Mars at an unprecedented resolution and in colour.

NASA switched off Viking 2 in July 1978 after a leak developed in its attitude control propellant system, but Viking 1 continued to return data for almost 1,500 orbits before activities were terminated in August 1980. This longevity enabled it to monitor global seasonal variations over two local years. Meanwhile the space agency was developing the next mission.

In the past, pairs of spacecraft had been dispatched to provide redundancy against the loss of a vehicle at launch or in transit to its destination. In the mid-1980s however, it was decided that the technology had improved sufficiently to justify sending out missions singly. To further reduce costs, these would exploit hardware and electronics employed by terrestrial communications and weather satellites.

The first such mission, Mars Observer, was launched in 1992 with state of the art sensors to conduct a survey of the surface and atmosphere of the planet over the course of a local year. In addition to a high-resolution imaging system it had a laser to map the topography; gamma-ray and thermal emission spectrometers to chart the composition of the surface; an infrared radiometer to monitor the cycles of the atmosphere; and an integrated magnetometer and electron reflectometer to study how the planet interacted with the solar wind.

Unfortunately, contact with the spacecraft was lost on 22 August 1993, just three days before it was scheduled to fire its engine to enter orbit around Mars. Since contact was never regained, the investigation was unable to identify the problem conclusively but it suggested that a slow propellant leak which would have been inconsequential on a satellite that used its main propulsion system only for a few days in manoeuvring to its operating point, had built up during the interplanetary cruise and caused an explosion when pyrotechnic valves were fired to open the feed lines in preparation for the orbital insertion burn. In addition, in retrospect it was acknowledged that because Mars is much farther from the Sun, the thermal environment would have been very different to that envisaged when that system was designed. In view of this disaster, the idea of reusing systems developed for Earth satellites was abandoned.

The new Mars Exploration Program was to exploit every launch window over the period of a decade to send a succession of probes, each carrying only a few instruments for a specific investigation. In this way, all of the instruments created for the lost Mars Observer would eventually be flown.

The three main themes were to understand the geology of the surface and subsurface, to understand the history of the climate, particularly the volatiles in the atmosphere, and to seek evidence of past and present life. As a search for water and water-modified geological features would address all themes, it was called the 'follow the water' strategy.

# Global Surveyor

The first of these new missions, Mars Global Surveyor, arrived on 12 September 1997 with the objective of providing data for at least one local year using the high-resolution camera, laser altimeter, thermal emission spectrometer, and integrated magnetometer and electron reflectometer inherited from Mars Observer.

It was to employ a similar Sun-synchronous mapping orbit to that planned for Mars Observer but attain this in a different manner. Whereas its lost predecessor would have fired its engine a number of times to manoeuvre into the desired orbit over a period of three months, the new vehicle was to use its engine to enter an initial 'capture' orbit that had a high apoapsis and, on reaching that point, fire its engine again to dip the periapsis into the upper atmosphere in order to use 'aerobraking' to gradually lower the apoapsis over a six-month period, whereupon the periapsis would be lifted out of the atmosphere. The rationale for this was to reduce the mass of propellant needed to attain the mapping orbit, and thereby enable the mission to be launched using a smaller, cheaper rocket. If it was successful, aerobraking was to become standard operating procedure.

But aerobraking requires caution as the drag force depends on the density of the upper atmosphere, which can vary with time, location, and solar activity. The altitude chosen

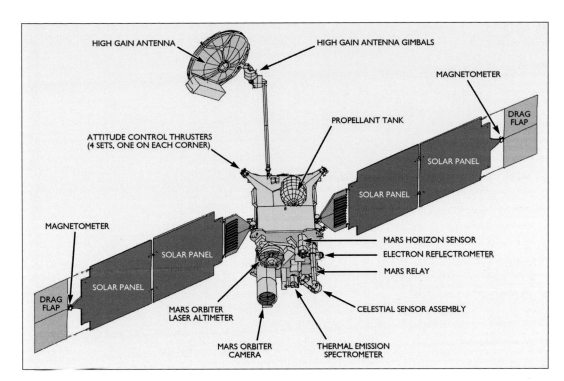

**LEFT** Details of Mars Global Surveyor. (NASA/JPL-Caltech/Woods)

for periapsis must balance a deep penetration of the atmosphere for rapid braking versus the ability of the vehicle to tolerate the resulting structural and thermal loads.

Mars Global Surveyor's two solar arrays were to cant at an angle of 30° with their rear surfaces facing in the direction of travel during aerobraking, in order to provide 'weathercock' stability. However, one them had failed to latch into place and it was reckoned the triangular epoxy-aluminium honeycomb that connected the panel to its gimbal was tearing free. The loose panel was rotated 180° in the hope that the force imparted during the first penetration of the atmosphere would cause it to latch into position. Once the panel was locked into place, it was to be rotated back to the planned angle for further aerobraking.

With the spacecraft in an orbit of 110 x 54,000km, the first aerobraking pass on 6 October imposed deceleration loads 50% greater than expected. This bent the damaged array beyond its nominal deployed position. Controllers instructed the spacecraft to lift its periapsis out of the atmosphere while they conducted an investigation. This concluded that the panel could tolerate a dynamic pressure of only one-third of that planned. It was necessary to carry out the aerobraking at a higher altitude in order to reduce the drag, at the expense of taking considerably

longer to achieve the circular mapping orbit at an altitude of about 375km, and in fact this was not attained until March 1999. The final orbit was chosen so that all images taken of a given surface feature on different passes were under identical lighting conditions; in effect, it was always early afternoon.

The imaging resolution of the camera on Mars Global Surveyor was so much better than that of the Viking orbiters that it dramatically improved our knowledge of the planet.

## Two hemispheres

The impression of Mars created by the flybys during the 1960s was of an ancient cratered plain, but between them these spacecraft had imaged barely 10% of the surface. When Mariner 9 mapped the planet from orbit it revealed a considerably more diverse landscape. And then our understanding was improved by the colour imagery from the Viking orbiters.

Until Mars Global Surveyor arrived, our knowledge of elevations was gleaned from measuring the pressure at the surface. As the newcomer orbited around the planet its laser altimeter measured the distance from the spacecraft to the ground directly below. Over time, these altimetry tracks created a topographical map that revealed the large-scale

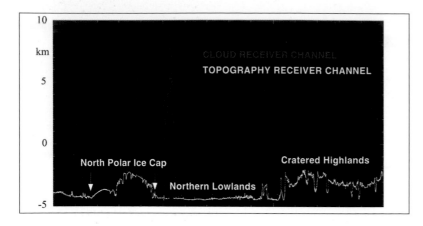

TOPOGRAPHY RECEIVER CHANNEL

North Polar Ice Cap

Cratered Highlands

Northern Lowlands

**LEFT** A single scan by the laser altimeter of Mars Global Surveyor as the spacecraft was aerobraking. The dichotomy between the low-lying northern plains and the southern highlands is clear. The instrument also made the first direct measurement of cloud heights, present at altitudes ranging from just above the surface to at least 10km. The clouds were observed primarily at the boundary of the ice cap and surrounding terrain. *(NASA/JPL-Caltech/GSFC)*

**RIGHT** Integrating altimetry data with precise radio tracking of Mars Global Surveyor enabled the topography of the planet to be correlated with its gravity field and crustal thickness. *(NASA/GSFC/Antonio Genova)*

**BELOW** A general topographic comparison of Earth and Mars (with different, unspecified colour scales). *(NASA/JPL-Caltech/GSFC)*

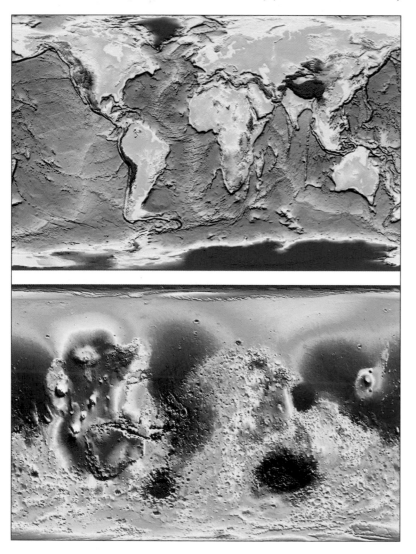

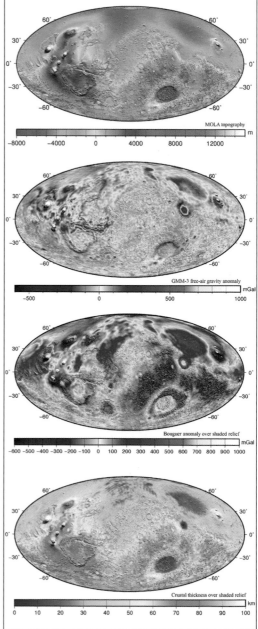

MOLA topography

GMM-3 free-air gravity anomaly

Bouguer anomaly over shaded relief

Crustal thickness over shaded relief

landforms in the same manner as terrestrial maps which include the ocean floors.

Generally speaking Mars has two morphologically distinct hemispheres. The most northerly point of the somewhat irregular transitional 'line of dichotomy' is at 330° of longitude, at 50° north. Significantly the terrain to the north of the line lies several kilometres below the 'datum' that is defined as the elevation at which the atmospheric pressure is the 'triple point' for water, namely 6.2 millibars.

The smooth plains of Vastitas Borealis may be either effusive volcanism or some other form of sedimentation on a vast scale. This region is comparatively lightly cratered, but the rims that protrude in places reveal that the infill material masks a low-lying cratered terrain.

The boundary is an irregular but fairly shallow scarp whose line is scalloped where major impacts created basins that were later breached from the north and flooded.

The huge volcanoes and flood channels suggest that early in its history the planet had a dense atmosphere and was warm and wet.

**BELOW** By integrating altimetry scans from many orbits the USGS produced a global topographic map, catalogue number I-2782. *(NASA/USGS)*

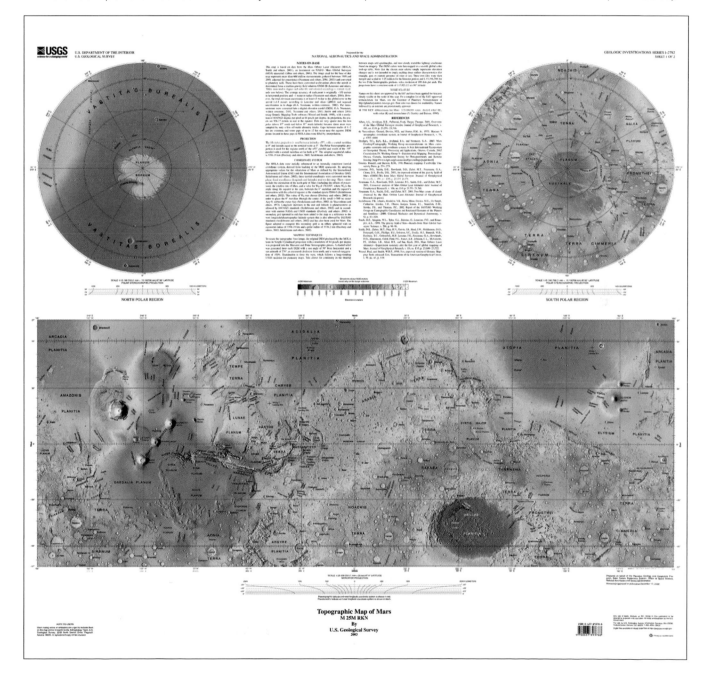

## Volcanic provinces

The province of Tharsis has an asymmetric outline 4,000km north to south and 3,000km east to west. Its steep northwestern flank rises from the northern plains and forms a series of young lava flows, but the shallow eastern flank transitions into the southern highlands.

To the north of this vast bulge lies Alba Patera, which was the first volcano to develop in this region.

Although the base of Alba Patera spans 1,500km, it is an extremely shallow edifice whose summit rises only 1km or so. Concentric fractures at the periphery indicate that it is in an advanced state of collapse. Early in its history the volcano may well have explosively issued clouds of ash in hot gas, a phenomenon known as a pyroclastic flow. In the low gravity, such a cloud would have been blasted to a height of 100km before collapsing and surging across the surface to produce a blanket of ash that formed a ring between 250 and 450km from the caldera. The style of eruption later became effusive, and sheets of lava were fed by channels and tubes which can be traced up to 1,000km west from vents near the irregular caldera complex that spans 100km.

Syria Planum lies 8 to 10km above the datum, marking the broad summit of Tharsis. It is intensely incised by the intersecting canyons of Noctis Labyrinthus. There are long fractures radiating down the flanks and others forming concentric arcs on the lower slopes. Deep crustal faults on the eastern flank have made the canyon system of Valles Marineris which extends one-fifth of the way around the equator.

On the northwestern flank of Tharsis lies a line of shield volcanoes spaced about 700km apart. These are the three 'spots' that Mariner 9 saw poking out of the dust storm. Southwest to northeast they are Arsia Mons, Pavonis Mons, and Ascraeus Mons.

Lunae Planum to the east of Tharsis is a lightly cratered plain characterised by widely spaced low ridges. Such plains are believed to be lava flows far more extensive than those produced by central vents. In that they were vast eruptions of low-viscosity lavas, they resemble the dark maria of the Moon. Like terrestrial 'flood basalts' such as the Deccan Traps and Columbia River Plateau they were probably erupted from fractures and left no trace of their sources. The plains are sufficiently thick to have completely masked the earlier topography. The 'wrinkle ridges' were made by compressional stresses as the lava cooled, densified and settled. The fact that many of the ridges on Lunae Planum are circumferential to Tharsis suggests that these stresses arose because the plain is now on a slight slope.

In an early image of the wall of Tithonium Chasma in Valles Marineris, Mars Global Surveyor saw a 1km high cliff that had 80 layers in outcrop. As imagery from other sections of the canyon system was obtained, it was realised that this stratification is present across a broad area. Since there has evidently been no lithospheric recycling on Mars, this layering may well represent a cross-section dating back to the earliest times.

To the northwest of Tharsis lies Olympus Mons, which is the largest volcano on the planet. Its lower flank is truncated by a scarp that matches the elevation of Syria Planum. The edifice is a vast stack of lava flows, some of which, particularly in the northeast and the

**BELOW A chart of the crustal faults in the Tharsis region which contains not only Olympus Mons and the line of three major stratovolcanoes but also the vast canyon system named Valles Marineris.** *(USGS/M.H. Carr)*

southwest, are draped over the scarp; in other places the scarp is an almost sheer cliff, but it follows an irregular line and in short sections it is oriented radially. Measured in terms of its peripheral scarp, the structure spans 550km.

The incline above the scarp is only a few degrees, but about one-third of the way up there is a succession of 10° ramps onto wide terraces that were probably formed by slippage on shallow thrust faults as the edifice inflated and deflated in response to magma rising into and draining from the upper magma reservoir. The summit stands 17km above the crest of the scarp. The caldera complex spanning 80km is marked by a cliff that plunges 2km to the multifaceted floor that indicates at least six phases of activity.

There are large lobate features located several hundred kilometres out from the scarp. The ridges and grooves imply deposits several kilometres thick. It has been suggested they are the material that slumped off Olympus Mons in a 'base surge' and exposed the scarp.

As the three large shields lie atop the Tharsis bulge at an elevation of 10km, their summits are 27km above the datum. The summit of Olympus Mons is at this elevation, but because it stands on terrain that is just 2km above the datum it is a larger overall structure.

Magma rising in a volcano is driven by the hydrostatic pressure induced by the difference in density between the magma and the rock it must pass through. The fact that all of the large shields peaked at the same elevation suggests that growth from their summit vents ceased when the pressure inside the planet was no longer able to force magma up the feed pipes.

The most instructive terrestrial comparison is the Hawaiian island chain. As the lithospheric plate that forms the floor of the Pacific Ocean migrates across a mantle 'hot spot', magma sometimes rises from a depth of 60km to make a series of volcanoes that form a line thousands of kilometres in length. Even if all of this magma had been limited to a single edifice it would not rival Olympus Mons. The tremendous bulk of the shield volcanoes of Tharsis is therefore evidence that the lithosphere of Mars is immobile.

To the west of Tharsis is Apollinaris Patera. This is in the transitionary terrain just north of the line of

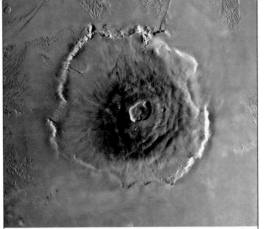

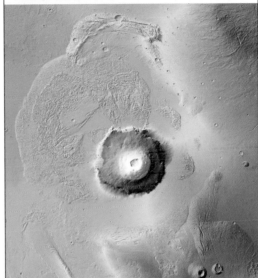

**LEFT** The lobate peripheral structures of Olympus Mons may be material that broke away when the 8km tall scarp formed around the edifice. The visual view is Viking orbiter imagery and the wider false-colour view is altimetry from Mars Global Surveyor.

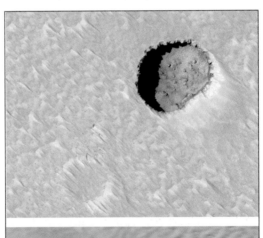

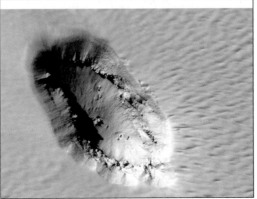

**LEFT** Pits on the flank of Arsia Mons, probably formed by the collapse of lava tubes. *(NASA/JPL-Caltech/Univ. of Arizona)*

dichotomy. It may have initially issued ash, but is believed to be the earliest of the lava shields.

And to the northwest of that, in the Elysium province, there are three shields that rise 5km above the low-lying northern plain.

### Volcanism in the highlands

At 2,000km in diameter, Hellas is the largest of the well-preserved impact basins. It has an arc of mountains extending to the north and west. The southeastern rim is degraded, the southwestern rim is masked by volcanism, and the northeastern rim is etched by two channels that deposited material on the floor. Nevertheless, the floor is 7km below the datum. The peripheral ejecta blanket, standing 3.3km above the datum, is the highest terrain in the southern hemisphere.

The reduction in pressure on the lower lithosphere by the removal of crustal material during such a large impact would have stimulated localised melting that rose through deep faults to the surface to create extremely low-profile structures with complex calderas.

Tyrrhena Patera lies 1,500km northeast of Hellas. A succession of explosive eruptions imply that as the magma neared the surface it came into contact with subterranean water that flashed to steam. The flanks of the shallow edifice are deeply eroded with radial channels, suggesting that pyroclastic flows left thick blankets of ash that later became 'welded' firm.

The complex caldera reveals episodic volcanism, and it has been suggested that once the vent exhausted its supply of volatiles it switched to effusive magma and that it was this, rather than water, that eroded the deep channels. One of the most prominent channels emerges from the caldera, extends west for more than 200km, and merges with the surrounding volcanic plain.

Hadriaca Patera, nearby, also underwent explosive eruptions that deposited thick blankets of ash. Its subdued caldera produced an apron 300km across that was so thoroughly etched by radiating channels that it is now distinctively ridged. Given the location of Hadriaca on the regional slope, its lava produced channels 400km long that drained to the southwest.

The nearest terrestrial analogue for the explosive phase of a Martian patera is Yellowstone in Wyoming, a 'resurgent caldera' that has repeatedly blanketed a large part of the North American continent with ash.

With flanking slopes of just a fraction of a degree, Amphitrites and Peneus Paterae to the southwest of Hellas are so subdued that their calderas resemble impacts on open ground, but their effusive lavas embayed the ancient cratered terrain.

Elsewhere, Nili and Meroe Paterae on Syrtis Major Planum portray another type of highland volcanism. They resemble ultra-low-profile shields on a volcanic plateau and exceed 1,000km in width. This volcanism is evidently related to the nearby Isidis basin in the same manner that Tyrrhena Patera is related to Hellas, because the vents lie on arcuate fractures concentric to the basin rim.

The fact that there are no low-profile central-vent structures on the low-lying northern plains implies the conditions that led to their formation only occurred in the ancient cratered terrain.

## Ancient magnetism

One advantage came from Mars Global Surveyor having to spend a longer-than-planned time in aerobraking, because when the spacecraft was at periapsis, the magnetometer was better able to sense the surface than it would in the planned mapping orbit above the ionosphere.

Today Mars does not have a global magnetic field but Mars Global Surveyor was able to detect residual magnetism in surface rocks.

**BELOW Radial magnetic field measurements by the electron reflectometer of Mars Global Surveyor indicated anomalies in crustal magnetism that evidently predate the formation of the large-scale surface structures.** (NASA/JPL-Caltech/GSFC/ J.E.P. Connerney)

MARS CRUSTAL MAGNETISM    $\Delta B_r$    MARS GLOBAL SURVEYOR    MAG/ER

UTOPIA PLANITIA · ARCADIA · ALBA PATERA · ACIDALIA · ELYSIUM MONS · CERBERUS RUPES · AMAZONIS · OLYMPUS MONS · CHRYSE · ISIDIS · THARSIS MONTES · VALLES MARINERIS · HELLAS · ARGYRE

East Longitude

$\Delta B_r / \Delta Lat$ (nT/deg)

-30 -10 -3 -1 -.3          +.3  1  3  10  30

When hot magma cools, it adopts the ambient field. On Earth, the north and south magnetic poles switch positions from time to time and the variation in this field can be mapped in rocks. In particular, where the crust spreads at ocean ridges there is a striped pattern in the solidified magma that is symmetrical on either side of the centreline; in effect, this is a tape recorder.

The magnetic stripes detected by Mars Global Surveyor are strongest at 180° longitude and 60° south, in the southern highlands. They are absent from the low-lying northern plains, from the volcanoes of Tharsis, Olympus Mons, and Elysium, and from the major basins such as Hellas.

Intriguingly, although the magnetism is associated with the oldest part of the surface, namely the cratered southern highlands, there is no correlation between the stripes and the individual landforms. This suggests the crust was magnetised very early on, before the bombardment that cratered the planet. The subsequent events which produced the volcanic provinces and filled in the low-lying northern region buried the early magnetised surface. Furthermore, the fact that these new structures are not magnetised indicates that by then the initially intense magnetic field had weakened.

## Valley networks

There are several types of valley network in the southern highlands. The runoff channels so resemble terrestrial drainage systems that the case for their having been etched by slowly running surface water is compelling.

These channels are typically less than 1km wide, seldom exceed 100km in length, start off small, and progressively grow by drawing upon dendritic feeder channels. In many cases they terminate abruptly, as if the surface water went underground at that point. On Earth, the erosion that forms karst in permeable rock often produces such 'blind alleys'.

There are also structures that look as if they were formed by collapse when permafrost melted and water erupted from the ground. Such valleys are more or less straight and possess steep walls. The tributaries come from amphitheatre-like cavities that show no indication of having served as runoff collectors. They are very short, join high on the walls, and

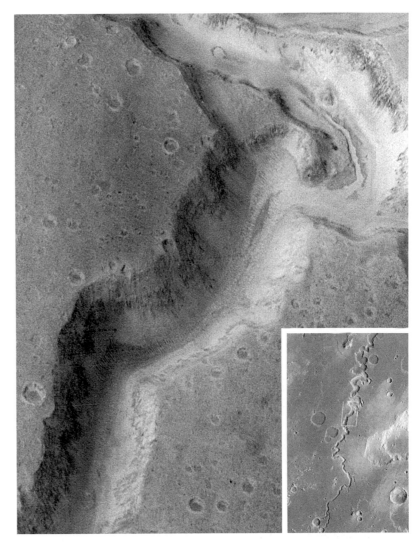

would have discharged into the main channel by waterfalls.

As an integral feature of the southern highlands, the various valley networks clearly derive from an early era, possibly when the atmosphere was pumped up by greenhouse gases emitted by the volcanoes and the climate was warmer and wetter.

## Flows in gullies

Early on, Mars Global Surveyor spotted indications that liquid water might have flowed on the surface in the geologically recent past.

There were gullies etched into the walls of some of the craters and channels in the southern hemisphere that appeared to indicate ground water seepage and runoff.

It was tempting to assume that they were etched by running water, but many of the locations were far from the equator, where

**ABOVE** The superb camera of Mars Global Surveyor resolved detail that was not evident in Viking imagery. Here, one of the meandering canyons of the Nanedi Valles system in the Xanthe Terra region of Mars has a narrow channel on its floor that looks like it was cut by a sustained flow of water. (NASA/JPL-Caltech/MSSS)

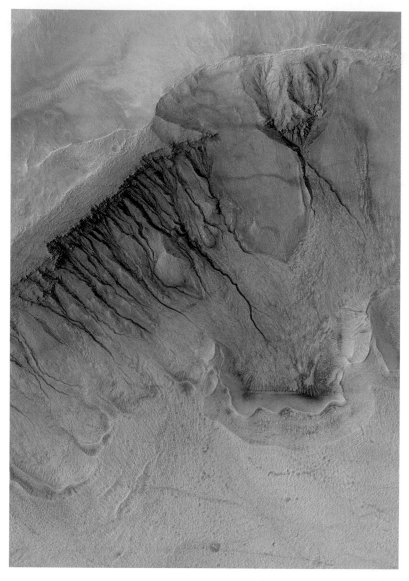

**LEFT** This view of the wall of the crater Newton in the southern highlands spans 1.5km. It was tempting to interpret the 'gullies' as fluid which leaked from the walls and flowed down into the crater. *(NASA/JPL-Caltech/MSSS)*

**BELOW LEFT** There are also streaks on slopes, as shown here inside a crater on Arabia Terra imaged by the even more powerful HiRISE camera of Mars Reconnaissance Orbiter. Believed to be the result of shallow cascades of dust, the streaks are dark when fresh and fade as they age. *(NASA/JPL-Caltech/Univ. of Arizona)*

the ground seemed likely to be frozen to a depth of several kilometres, whereas the gully sources are only a few hundred metres downslope. Nevertheless, if water was being driven towards the surface and bursting from strata exposed in steep slopes, the distribution of sites implied regional subterranean aquifers. Perhaps geothermal energy sporadically melted the underside of the permafrost to create these aquifers.

But there are rival hypotheses, one being that the gullies were etched by the release of gases present in ice. If carbon dioxide was vented from a permafrost exposure, it would contain about 1% of water by volume and this would create a muddy apron downslope. Unfortunately the spatial resolution of the spacecraft's thermal imaging spectrometer was insufficient for it to differentiate the chemical composition of these small aprons from their surroundings.

Later missions would further investigate the process by which these gullies formed.

## Impact craters

As the highlands are not 'saturated' with craters, there are spaces between them. Because the intercrater material masks ejecta blankets and encroaches on some of the craters themselves it must have been emplaced later. Most of this material is believed to be volcanic ash blasted from vents and lava erupted from fissures.

The ejecta blankets of some craters also indicate the past presence of water. As a rule, Martian craters do not possess extensive blankets of hummocky ejecta. On the Moon, the continuous ejecta deposit is typically confined within 0.7 radii of a crater's rim, but the continuous ejecta for craters on Mars extends out to 2 radii. Debris travelling on a ballistic trajectory would be expected to travel farther on the Moon than on Mars due to the gravity of the Moon being weaker. Why should the ejecta extend farther from craters on Mars? The answer would appear to involve the form of the ejecta. Ejecta on the Moon is usually blocky close to the rim and grades progressively to finer debris which ultimately blends into the surrounding terrain, but many craters on Mars exceeding 5km in diameter are surrounded by overlapping 'sheets' with lobate margins.

There is considerable variety in the form of such ejecta. In some cases, it is thin and the underlying topography is visible through it. In other cases, it masks the earlier terrain. Sometimes the ejecta is concentrated in a thick annulus, and usually there is a distinct radial pattern. Evidently these impacts 'splashed out' a blanket of slushy ejecta which then flowed over the surface. In some cases, the mudflow has either been deflected around a nearby crater's rim or sloshed over and flooded it.

The variation in crater types may therefore indicate that the excavated rock had differing amounts of water and ice. There is a geographical correlation, with the fluidised ejecta running farther from the rim of craters at higher latitudes.

## Outflow channels

Another surprising revelation by Mariner 9 was what appeared to be vast outflow channels. Apart from two on the eastern rim of the Hellas basin and some on the flank of the Elysium rise, these channels drain from the southern highlands onto the low-lying northern plains.

In fact most of the outflow channels cross the line of dichotomy and debouch onto Chryse Planitia. This vast drainage system comprises Ares, Tiu, and Simud (all of which emerge from collapse structures, typically 100km across, known as 'chaotic' terrain), Shalbatana (which derives from a smaller chaotic zone north of Eos), Maja (which is drawn from Juventae Chasma on the eastern side of Lunae Planum), and Kasei (which emerges from Echus Chasma located to the west of Lunae Planum and runs north prior to swinging east to drain into Chryse).

Although such channels emerge from their sources fully formed, some have short tributaries that emerge from their own chaotic terrains.

It is possible these sources formed on ring fractures that are concentric to the impact basin beneath Chryse Planitia. In this scenario the permafrost was melted by magma that exploited these deep fractures, and once the water found an outlet and drained the 'pore space', the roof collapsed into the hole to make the chaotic zone.

Contrariwise, it has been argued that the flooding was on such an enormous scale because the sudden 'breakout' of water onto the surface was 'pumped' by gravity-driven artesian flow inside the Tharsis bulge.

However they originated, these floods were extremely erosive. In particular, they swept away portions of the ejecta blankets surrounding craters and incised their flanks to leave them standing on 'pedestals'. If obstructed by a ridge, they accumulated until the water level reached the crest, then cut a deep channel as the dam drained. On debouching onto the low-lying plains, they spread out and deposited the suspended material, leaving gravel bars through which later flows eroded braided channels. And isolated obstacles on the open plain were turned into teardrops whose tails pointed downstream.

A close terrestrial equivalent of how such a landscape was carved is the Scablands of eastern Washington State. This was formed when an ice dam in northern Idaho broke at the end of the last Ice Age and, over time, released a large glacial lake equivalent to Lakes Erie and Ontario. The cataclysmic floods etched deep channels as they drained into the Pacific Ocean via the Columbia River Gorge.

## A northern ocean?

Conducting a detailed study of Viking orbital imagery in 1986, Timothy Parker of the University of Southern California found subtle features on the northern plains that bore a strong resemblance to the margins of a lake that had flooded parts of Utah, Nevada and Idaho some 100,000 years ago. After integrating the evidence, in 1989 he suggested

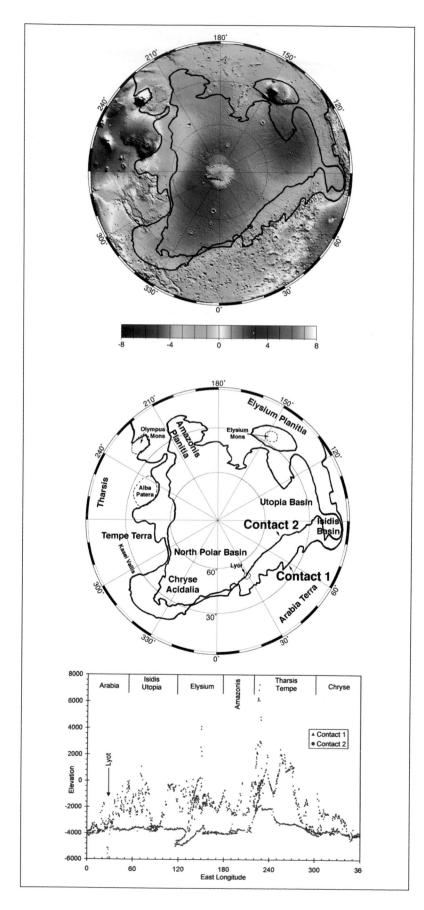

**LEFT** Two possible shorelines for an ocean that may have existed in the low-lying northern plains early in the history of Mars. Altimetry from Mars Global Surveyor showed the elevation of the inner contact to be remarkably constant apart from sections where there was clear evidence of subsequent geological activity, such as in the Tharsis region. *(NASA/MOLA Team/J.W. Head)*

that a series of oceans with differing amounts of water had formed and retreated north of the line of dichotomy. It was speculated that these oceans formed when magma intrusions melted permafrost.

The laser altimeter on Mars Global Surveyor could measure elevations to an accuracy of 1m, and it revealed that the elevation of the northern plains varies by no more than about 100m over distances of hundreds of kilometres. The profiles along the line of dichotomy were reminiscent of the continental slopes that sweep down onto the abyssal sedimentary plains of terrestrial oceans.

At an elevation 2km below the datum, the larger of two shorelines proposed by Parker, which he called the Arabian shoreline, traces an undulating elevation. But the entire circumference of the inner Deuteronilus shoreline, which is 3.5km below the datum, traces a single contour remarkably well, deviating by no more than 280m from its mean, and where the 'fit' is poor the reason is apparent in terms of subsequent tectonism and lava flows.

To James W. Head of Brown University, Rhode Island, the altimetry argued for an ocean having once existed within the inner shoreline. Its margin traces an almost level surface, and the topography inside it is more subdued than outside, which is consistent with smoothing by sedimentation. This has been tentatively named Oceanus Borealis.

When Mars Global Surveyor returned high-resolution imagery of parts of the putative shoreline there was little on the surface to support this contention. But it could be that on a planet dominated by aeolian erosion, an ancient shoreline will be so eroded as to be difficult to discern close up, and hence is best inferred from afar.

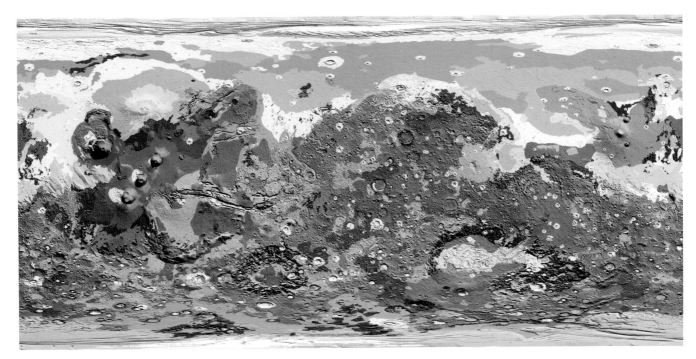

For example, Head has observed that over a distance of 2,200km six major outflow channels debouched onto Chryse within 180m of the elevation matching the Deuteronilus shoreline.

A later study of the highest-resolution altimetry established that all ten large channels that drain north across the line of dichotomy terminate between the two putative shorelines, with even the largest channel having lost its morphology by the inner one.

Head concluded from the terminations of the channels that drain into Chryse that they entered an ocean whose depth was 600m on average and 1,500m at its deepest point. This volume is consistent with estimates of the amount of water on the planet.

### Warm and wet to cold and arid

The morphological evidence for the previous existence of surface water on Mars strongly suggested that the early atmosphere was able to sustain a hydrological cycle which eroded the surface both physically and chemically. If intense rainfall drained into an ocean for a prolonged time, then by terrestrial analogy the bed of this ocean must be lined with carbonate sediments.

The main job of the thermal emission spectrometer on Mars Global Surveyor was to map the sediments and evaporites that would have been deposited in this putative warm and wet era. In fact it found no trace of such minerals. However, it did find olivine, a silicate present in igneous rock, to be commonplace, and since this mineral readily weathers in a warm and wet environment its presence meant the planet must have been cold and dry since the olivine was formed, which was presumably billions of years ago! This contradiction represented a considerable dilemma.

On 2 November 2006 routine commands were sent to Mars Global Surveyor. Sadly, something went wrong and it was never heard from again. But the 'low-cost' spacecraft had exceeded its design life by a factor of five.

## Feasts for the ghoul

Having had no luck at Mars in the early 1970s, the Soviet Union tried again more than a decade later by sending a pair of sophisticated spacecraft, Fobos 1 and 2, on an ambitious mission to orbit Mars and examine Phobos, the larger of its two small moons.

The original concept was for the spacecraft to land on Phobos, but its weak and irregular gravitational field would make this task extremely difficult. Next the planners considered having the vehicle manoeuvre to within 20m of the moon in order to collect samples by using harpoons. On deeming this operation to be too risky, they decided on deploying several small landers during a close flyby. The first task was to achieve

**ABOVE The NASA Planetary Geodynamics Laboratory used the altimetry to create a geological map.** (NASA/GSFC)

control thrusters at a moment when the vehicle was not in communication with Earth; thus nobody was aware of the erroneous response when the vehicle began to rotate and rapidly lost solar power. It was never heard from again. It was a fault that ought to have been detected on the ground, prior to uplinking the sequence.

Fobos 2 entered orbit around Mars on 30 January 1989 but was lost on 27 March while making a manoeuvre in close proximity to its objective, Phobos. In fact these spacecraft had been doomed prior to launch, as their computers had components that were known to be unreliable. One of the processors on board Fobos 2 failed in interplanetary space and a second began to suffer intermittent malfunctions soon after the vehicle entered orbit around Mars. Since the 'voting logic' was not reprogrammable, it was impossible to prevent the two faulty units from overriding the surviving third processor.

Financial constraints following the collapse of the Soviet Union prevented Russia from developing a new mission to Mars until 1996, when it launched a sophisticated spacecraft to study many aspects of the planet using a payload supplied by international teams. Unfortunately, Mars 8 was stranded in Earth parking orbit.

The NASA Mars Exploration Program called for launching missions at each opposition.

In addition to the infrared radiometer inherited from Mars Observer to profile the atmosphere, Mars Climate Orbiter had a new colour-imaging system capable of taking horizon-to-horizon images at medium resolution. It was to be a Martian weather satellite monitoring clouds and surface frost, and the daily and seasonal variations of vapour.

The capture orbit on arrival in September 1999 would not be as eccentric as that of Mars Global Surveyor, to enable it to rapidly use aerobraking to attain the desired Sun-synchronous orbit. It was to monitor the polar caps and atmosphere for one Martian year to gather data that it was hoped would help to shed light on whether the planet had undergone cyclic changes in climate over approximately the last 100,000 years.

Unfortunately the spacecraft was lost while firing its engine behind the planet to enter the capture orbit. The investigation found a simple

orbit around Mars in the same plane as the target, then undertake a series of manoeuvres to set up the desired flyby geometry.

This was a major enterprise, with many countries from both sides of the Iron Curtain participating, including the European Space Agency. American scientists joined some of the teams in secondary roles.

Unfortunately, the computer technology available to the developers was not up to the job.

The two vehicles set off in July 1988, but a commanding error led to the loss of Fobos 1 in August. A fault in the sequence caused the spacecraft's computer to deactivate the attitude

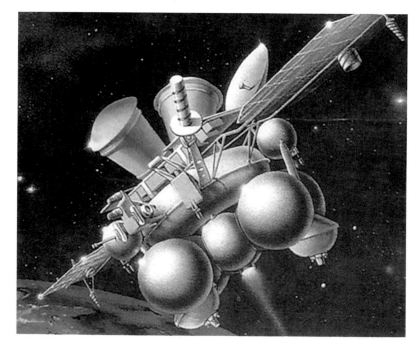

unit-conversion error. A data file sent from the contractor to the JPL engineers who were to control the trajectory was supposed to have been expressed in metric units but was instead in Imperial (so-called 'English') units. As a result, the spacecraft flew too close to the planet. The first sign of trouble came when it slipped behind the planet's limb 49sec earlier than predicted. When the trajectory was recalculated by taking into account the unit-conversion error, it was realised the vehicle had penetrated the atmosphere to an altitude of 57km, causing it to burn up like a meteor.

As a result, NASA revised procedures to preclude a repeat of the confusion about units of measurement, and set more conservative margins for the orbital insertion and aerobraking manoeuvres.

When Japan decided to participate in the exploration of Mars it designed an orbiter called Nozomi (Hope) to investigate the ionosphere and how it interacted with the solar wind.

As Japan did not have a rocket sufficiently powerful to send the spacecraft to Mars, when it was inserted in 1998 into a highly elliptical Earth orbit the plan was to raise the apogee using two flybys of the Moon prior to a propulsive manoeuvre to escape. Unfortunately, the propulsive system sprang a leak during that escape burn and the trajectory left the spacecraft in a heliocentric orbit that fell short of its target. In a revised plan, a series of Earth flybys set up an approach to the planet in December 2003. However, efforts at that time to orientate the vehicle to fire its engine for orbit insertion failed and it made a 1,000km flyby.

When Russia launched its Fobos-Grunt (Soil) mission in November 2011, the spacecraft had a Chinese probe named Yinghuo (Firefly) as a piggyback payload.

The main objective of Fobos-Grunt was to deploy a probe to land on Phobos and collect a sample of about 200 grams of loose soil that would then be returned to Earth for analysis. If the Martian moon was a captured asteroid then this would be a welcome sample of such a body. If, however, the moon was more intimately related to Mars, the mission would indirectly provide a sample of the planet itself.

Meanwhile, Yinghuo would remain in the highly elliptical capture orbit with an apoapsis

of 80,000km to study interactions between the solar wind and the upper atmosphere.

However, as had happened to Mars 8 in 1996, the mission was stranded in Earth parking orbit.

Project planners had been jokingly wondering for some time whether Mars was cursed. The rumours of a 'ghoul' began in 1969 when Mariner 7 briefly fell silent shortly prior to its flyby of the planet.

## Odyssey insights

As there had been nothing fundamentally wrong with the Mars Climate Orbiter vehicle, just how it was operated, NASA was able to send a replacement at the next launch window. The fact that this was in 2001 led to it being named Mars Odyssey, in homage to the novel *2001: A Space Odyssey* written by Arthur C. Clarke.

It entered orbit around Mars on 24 October 2001 and began aerobraking. By January 2002 it had achieved a circular, Sun-synchronous orbit and set about its primary mission.

It carried two instruments. A gamma-ray spectrometer (the final inheritance from Mars Observer) would measure the abundances of certain elements in the topsoil at a spatial resolution of 300km to provide a sense of the character of the surface. This instrument was now augmented with a neutron detector capable of sensing hydrogen in the uppermost metre or so of the ground, as an indicator of either

**ABOVE** A depiction by G. W. Burton of a malevolent creature interfering with Mariner 7 as it approaches Mars in 1969, with a mark of puzzlement above the distant Earth. It would feast many times in future decades. *(JPL-Caltech)*

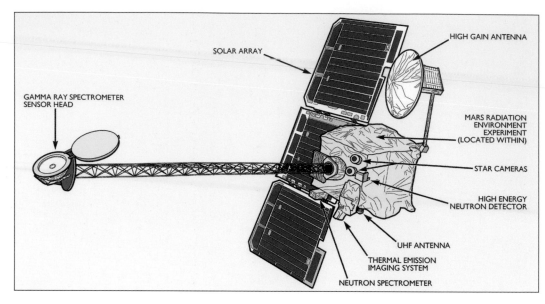

**BELOW** The neutron detector of Mars Odyssey mapped the abundance of hydrogen in the upper metre or so of the Martian surface, showing the distribution of hydrated minerals and/or a water ice permafrost. (NASA/JPL-Caltech/LANL/Univ. of Arizona)

hydrated minerals or water ice permafrost. And there was an entirely new instrument, a thermal emission spectrometer that was integrated with a medium-resolution optical imager. This instrument was capable of detecting carbonates, silicates, hydroxides, sulphates, oxides, and hydrothermal silica in the topsoil. The plan was to create a global mineralogical map at a resolution of 100m/pixel and correlate this with the visible evidence, to identify structures associated with the action of water that could yield insight into the past climate.

As the first orbiter capable of detecting near-surface ice away from the polar regions, Mars Odyssey was expected to significantly advance our knowledge of where the ancient surface water went.

As soon as the neutron detector started to operate it achieved the mission's most significant result.

It discovered that the abundance of hydrogen varied quite widely across the planet. Not entirely unexpectedly there were vast reserves of water ice close to the surface in the polar regions, probably a mix of ice and dirt called permafrost. What was surprising was hydrogen-rich terrain at mid-latitudes and some sites near to the equator where it was too warm for ice to be stable near the surface, raising the prospect that this water was chemically bound into minerals such as clays.

Further study established that the amount of water ice away from the polar regions was too great to be currently in equilibrium with the atmosphere, which suggested the planet was emerging from a colder period or 'ice age' that ended as recently as 400,000 years ago and that the ice deposits were still adapting to the new climate.

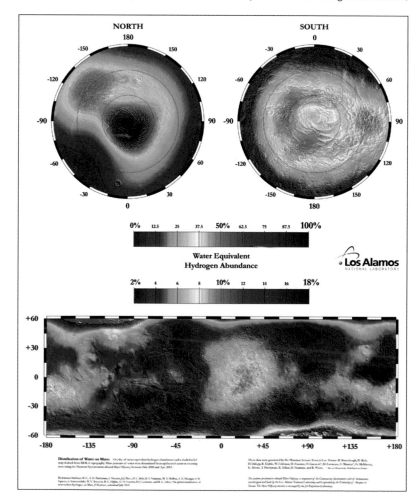

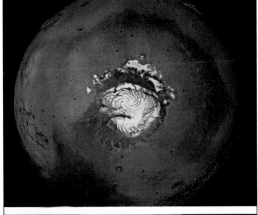

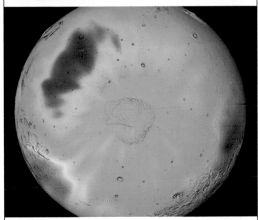

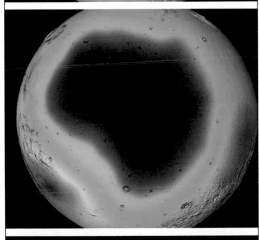

Counts/second

| | | | | | | | | | | | | | | | | |
| 0.0 | | | | | 0.1 | | | | | 0.2 | | | | | 0.3 |

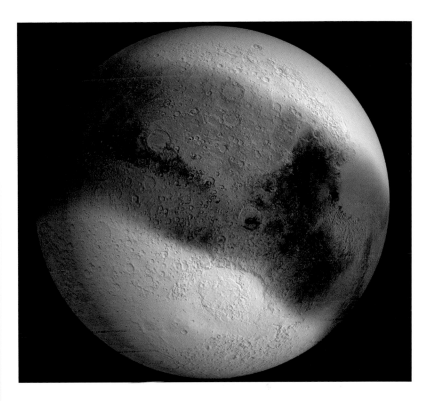

**ABOVE** The climate of Mars is sensitive to small changes in orbit eccentricity and planetary axial tilt. Models produced using Mars Global Surveyor and Mars Odyssey data suggest that a steeper tilt existed between about 2.1 million to 400,000 years ago and this increased solar heating at the poles. This polar warming would have caused mobilisation of water vapour and dust into the atmosphere, with the build-up of a surface deposit of ice and dust down to 30° of latitude in both hemispheres. By this reckoning, the planet has been in an interglacial period with its axis less tilted for the last 300,000 years or so, during which time the ice-rich surface deposit has been degrading in the latitude zone of 30° to 60° as water-ice returns to the poles. This graphic overlays those surface deposits on a topographic map based on altimetry by Mars Global Surveyor. *(NASA/JPL/Brown Univ.)*

**LEFT** A Viking orbiter view of the northern hemisphere provides context for Mars Odyssey data. In the middle image blue shows water ice during the northern winter, when the surface is covered with a seasonal cap of frozen carbon dioxide. The bottom image shows the water ice that becomes evident when the seasonal cap sublimates in the summer. *(NASA/JPL-Caltech/GSFC/IKI)*

In addition, computer simulations showed that the spin axis of the planet is unstable and that during periods when the axis is steeply inclined the polar caps will sublimate and ice will migrate to form glaciers several kilometres thick on the slopes of Olympus Mons and the volcanoes of the Tharsis bulge. Indeed, there are many surface structures suggestive of glaciation.

Long-term observations of the polar regions revealed the seasonal variation of the ice caps as carbon dioxide sublimated and water ice was exposed.

The case for Oceanus Borealis has ebbed and flowed over the years, with the same data often being interpreted both for and against it. The gamma-ray spectrometer provided the best evidence to date in the form of potassium, iron, and other elements that could have

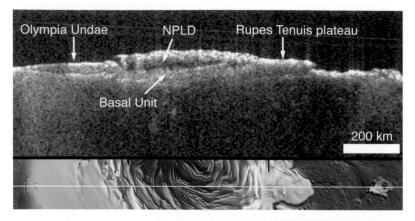

Olympia Undae    NPLD    Rupes Tenuis plateau

Basal Unit

200 km

**LEFT** A scan by the radar of Mars Express along a 15,000km track which passes over the ice-rich north polar plateau. It detected layers to a depth of almost 3km. The basal unit of a sand-rich and dust-rich icy material comprises more than half of the bulk of the polar plateau in this profile. Its base can be traced from beneath the Olympia Undae sand sea (on the left), across the entire polar stack to the margin of the Rupes Tenuis plateau where there are no overlying north polar layered deposits (NPLD). Note that the vertical dimension measures the time delay of the radio-signal echo. The deepening of the lower boundary of the basal unit at the centre is an artefact of the slowing of the radar wave in the icy material. In fact, this boundary is nearly flat. *(ESA/NASA/JPL-Caltech/Univ. of Rome/ASI/GSFC)*

These results were important for planning future surface missions, because clay deposits could represent some of the best places to seek evidence of past life. In particular, unlike sulphates which form in highly acidic water, clays form in watery conditions that are conducive to (terrestrial) life.

Since different environmental conditions were required for the production of sulphates and clays on Mars, the presence of both would

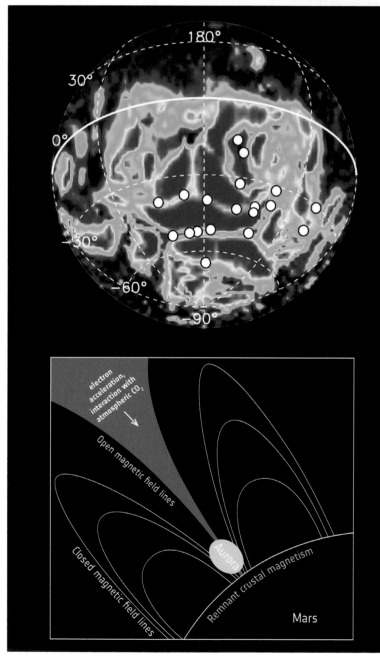

**LEFT** With a decade of data from Mars Express, scientists were able (for the first time) to combine all of the remote sensing observations of localised ultraviolet aurora with in-situ measurements of electrons striking the atmosphere. The plot shows the locations of 19 auroral detections (white circles) over the southern hemisphere during the night. The background portrays the magnetic field line structure as discovered by Mars Global Surveyor while it was aerobraking. Red indicates the closed magnetic field lines which are associated with relic crustal magnetism, grading through yellow, green, and blue to open field lines in purple. The auroral glows are very brief, they are not seen to repeat in the same locations, and they only occur close to the boundary between open and closed magnetic field lines (as shown in the diagram) in the altitude range 127±27km. *(ESA/ATG Medialab, J-C. Gérard & L. Soret)*

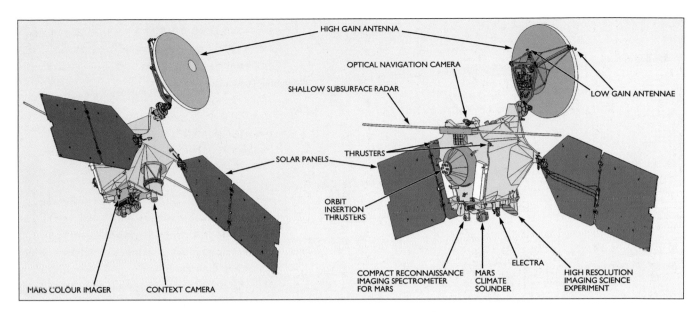

HIGH GAIN ANTENNA

OPTICAL NAVIGATION CAMERA

SHALLOW SUBSURFACE RADAR

LOW GAIN ANTENNAE

THRUSTERS

SOLAR PANELS

ORBIT
INSERTION
THRUSTERS

ELECTRA

MARS COLOUR IMAGER

CONTEXT CAMERA

COMPACT RECONNAISSANCE
IMAGING SPECTROMETER
FOR MARS

MARS
CLIMATE
SOUNDER

HIGH RESOLUTION
IMAGING SCIENCE
EXPERIMENT

appear to suggest two distinct water-related episodes separated in time, with the chemistry changing from neutral or alkaline water for the clays to acidic for the sulphates, perhaps in response to the Tharsis volcanism pumping sulphur into the atmosphere.

And, as already noted, because all the water-related deposits appear to be billions of years old, the surface has evidently been arid for most of its history.

Averaging data collected by the infrared spectrometer aboard Mars Express over many orbits established spectroscopic evidence of methane. The detection of methane in the atmosphere had been reported in 2003 by near-infrared high-resolution spectroscopy of terrestrial telescopes. It seemed to account for about 10 parts per billion of the already tenuous atmosphere on average but terrestrial monitoring had reported concentrations tens of times greater over certain areas. Mars Express confirmed that the methane was not uniformly distributed and that its abundance varied greatly over time.

As methane will be short-lived in such a highly oxidising atmosphere, for it to persist it must be being replenished. Over 90% of the methane in the terrestrial atmosphere is released by micro-organisms and the rest is of geochemical and volcanic origin. The Viking tests seemingly ruled out biology at the surface so if there were micro-organisms they would be in 'sheltered' environments beneath the surface, perhaps in hydrothermal springs – although infrared scans by Mars Odyssey hadn't found

such activity. Alternatively, the atmosphere might gain an occasional infusion of methane by the impact of a small cometary nucleus.

The origin of methane and other volatiles in the atmosphere of Mars became the subject of a later mission.

## Reconnaissance targets

On arriving in March 2006, Mars Reconnaissance Orbiter began aerobraking. It achieved its circular, Sun-synchronous operating orbit in September.

It had a super-high-resolution camera, an improved spectrometer to identify minerals at the surface, atmospheric instruments carried

**ABOVE** Details of **Mars Reconnaissance Orbiter.** *(NASA/JPL-Caltech/Woods)*

**BELOW** Preparing the high-resolution imaging science experiment (HiRISE) camera for Mars Reconnaissance Orbiter. *(NASA/JPL-Caltech/Ball Aerospace)*

## ADVANCED ATMOSPHERIC SOUNDING

The climate sounder for Mars Reconnaissance Orbiter was an improved model of an instrument on the lost Mars Climate Orbiter, which was itself a development of one on the lost Mars Observer. So it was third time lucky for the team!

It was a spectrometer with one visible/near-infrared (0.3 to 3.0μm) channel and eight far-infrared (12 to 50μm) channels. By viewing both down at the nadir and horizontally through the atmosphere to either side of the ground track it was able to quantify data in vertical slices of the atmosphere in increments of 5km to create daily global weather maps in terms of temperatures, pressures, humidity, and dust density.

**BELOW A secondary camera aboard Mars Reconnaissance Orbiter captures a global view of Mars and its weather on a daily basis. On Earth this data is mapped onto a 3D globe of the planet. This composite shows one such image for each month of 2014 (from left to right and top to bottom). Of particular interest are bright water-ice clouds which tend to cling to the summits of volcanoes, and the yellowish dust storms.** *(NASA/JPL-Caltech/MSSS/Bill Dunford)*

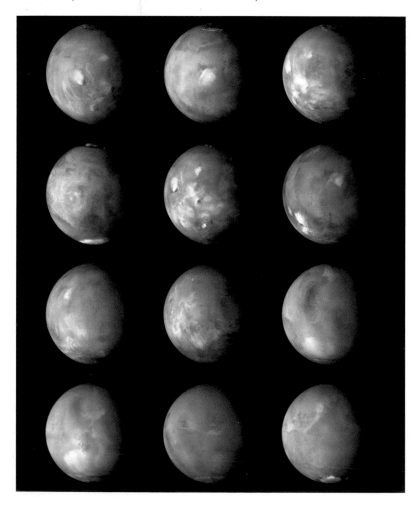

over from the lost Mars Climate Orbiter, and a high-resolution subsurface radar.

It was to monitor the weather and climate, and seek additional evidence of water ice. Specifically, it would examine water-modified rocks and ice deposits identified by Mars Odyssey and Mars Express and extrapolate the ground truth that was being provided by the Mars Exploration Rovers on the surface.

A very high-resolution analysis by Mars Reconnaissance Orbiter of Vastitas Borealis found no minerals suggestive of a deposit of very fine sediments on the floor of an ancient ocean. In fact it was littered with metre-sized rocks. So if there was once an ocean the sediments on its floor must have been buried. A study of the northern plains found some craters which showed spectral signs of clays and other hydrated minerals. The existence of buried hydrated minerals in ejecta was not in itself proof of ocean sediments, but it served to keep that hypothesis alive.

The ground-penetrating radar on Mars Express revealed circular structures ranging up to 470km wide buried beneath the smooth northern plains. There had been hints of them in Mars Global Surveyor laser altimetry as enigmatic circular depressions. Clearly, a tremendous amount of infill would have been required to mask such large structures. The material that currently forms the surface would depend upon how much volcanic activity occurred after the putative early ocean vanished.

The benefit of an orbital mission that lasted several Martian years was that it was able to monitor seasonal and annual cycles in the atmosphere and on the surface. An intriguing discovery by Mars Global Surveyor was changes in small gullies on slopes that gave the impression of material having flowed downslope and around obstacles, perhaps a salty fluid transporting small particles of sand. Surprisingly, however, most of the gullies were in ice-free locations.

This phenomenon was further investigated by Mars Reconnaissance Orbiter, whose infrared spectra did not show any evidence of the hydrated minerals that would be expected if salty groundwater were being released.

Laboratory studies have established that in the weak Martian gravity, which is 38% of Earth's, it would be feasible for

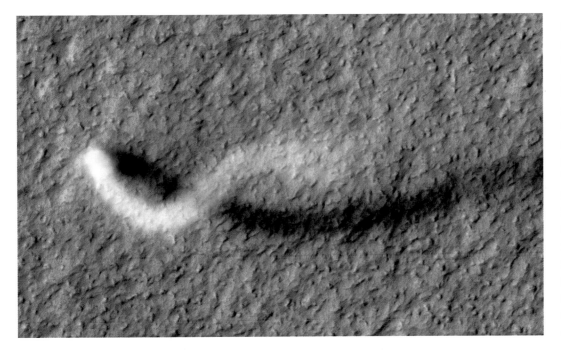

## ADVANCED SUBSURFACE RADAR

The radar carried by Mars Reconnaissance Orbiter was a derivative of the Mars Express instrument.

With a 10m antenna operating in frequency bands between 15 and 25MHz the radar was to distinguish layers as thin as 7m down to a depth of 1km with a vertical resolution of 10 to 20m. It was to be used primarily over the night side, where the ionosphere was weakest.

RIGHT This north polar scan by the radar of Mars Reconnaissance Orbiter detected four thick layers of ice and dust separated by almost pure ice. Calculations indicate that these thick ice-free layers represent a series of one-million-year-long cycles of climate change caused by variations in the planet's orbital eccentricity and axial tilt. Adding up the entire stack of ice therefore gives an estimated age for the north polar ice cap of about 4 million years. The image on the right shows the layered deposits and the basal unit in outcrop exposed near the edge of the cap. On the left is a map of the thickness of the layered deposits. At the bottom are the surface elevation of the polar region and the elevation at the base of the layered deposits. (NASA/JPL-Caltech/Univ. of Rome/SwRI/Univ. of Arizona)

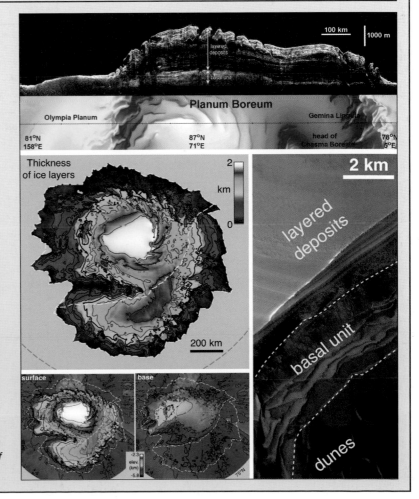

dry, unconsolidated granular debris to flow downslope and produce such features. This suggested that liquid water might not be involved after all.

Long-term studies of the appearance of new gullies indicated they were more likely to form during the winter, further suggesting that water was not involved, as that was the season in which water was *least* likely to melt.

Researchers instead suggested that the gullies were due to the winter accumulation of carbon dioxide frost triggering avalanches of dry sand.

Nevertheless clear signs of the presence of water ice in the upper crust were found in newly formed craters, where ice melted by the heat of impact could form channels, branched fans, and ponds.

Mars Reconnaissance Orbiter followed up on

**BELOW In 2014 the USGS produced this geological map of Mars, catalogue number SIM-3292.** *(USGS/NASA)*

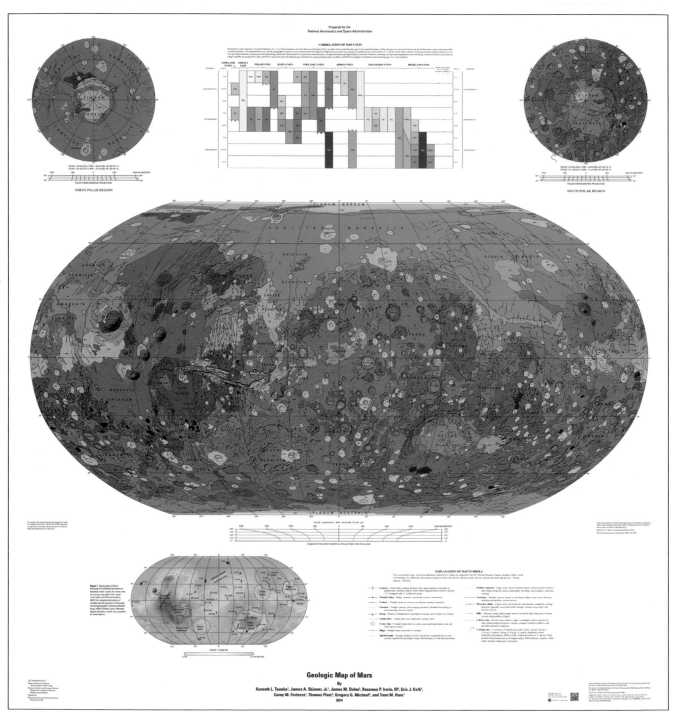

Geologic Map of Mars

By

Kenneth L. Tanaka[1], James A. Skinner, Jr.[1], James M. Dohm[2], Rossman P. Irwin, III[3], Eric J. Kolb[4],
Corey M. Fortezzo[1], Thomas Platz[5], Gregory G. Michael[5], and Trent M. Hare[1]

2014

the detection by Mars Odyssey of chloride salts in low-lying parts of the southern highlands, showing them to be sites where surface water had once pooled and later evaporated to leave behind crystals.

It also investigated freshly made impact craters several metres in diameter. A spectroscopic study of one that appeared in 2008 found its white ejecta blanket to be water ice. The blanket disappeared across several months, just as sublimating ice should do. Remarkably, if the shallow ice slab implied by these observations were also present beneath the Viking 2 site in Utopia Planitia at a similar latitude, then the scoop of that 1970s lander would have come within 10cm of exposing it!

In addition to studying the structure and cycles of the polar caps, the surface-penetrating radars on Mars Reconnaissance Orbiter and Mars Express revealed subsurface water ice to be common. However, they did not find any liquid water aquifers. If these exist they must be at depths beyond the reach of orbital radars.

## A geological timescale

The accumulated results of the recent orbital and surface missions support the hypothesis that abundant water had existed in the past on a warmer Mars. But scientists remain divided about the times and durations of these wetter periods.

A geological timescale has been devised for Mars based upon the character of its surface, the stratigraphic relationships between the various different units, and their ages as inferred from counting impact craters.

The oldest is called the Noachian, this name being derived from the heavily cratered highland region west of the Hellas basin. About 45% of the planet is of Noachian age. Next is the Hesperian, which saw widespread volcanism and the catastrophic flooding that made the immense outflow channels. It represents the transition from the wetter and perhaps warmer world of the Noachian to the arid, cold, and dusty Amazonian which persists today.

Of course, this timescale is relative. The absolute ages are uncertain, but it seems reasonable to say that the Amazonian began around 3 billion years ago, give or take a couple of hundred million years. In order to progress further it will be necessary to collect a representative suite of rocks and return them to Earth for dating by radiogenic methods.

## India's Mars Orbiter

If there exists a ghoul in space, it must have been asleep when India joined the exploration of Mars. Launched in November 2013, its Mangalyaan (Mars Craft) scored a remarkable success.

The spacecraft spent some time in Earth orbit gradually raising its apogee prior to the escape burn. On arrival at Mars in September 2014 it successfully entered a steeply inclined, highly elliptical orbit well-suited to studying how the solar wind erodes the planet's atmosphere.

In fact, science was a welcome bonus because the purpose of the mission was to test technologies for follow-on missions.

## Atmosphere leaking away

NASA's Mars Atmosphere and Volatile Evolution Mission (MAVEN) arrived in September 2014 and entered directly into a steeply inclined orbit that ranged between 150 and 6,200km with a suite of 'particles and fields' instruments to investigate interactions between the solar wind and the atmosphere.

**BELOW The MAVEN spacecraft.** *(NASA/GSFC)*

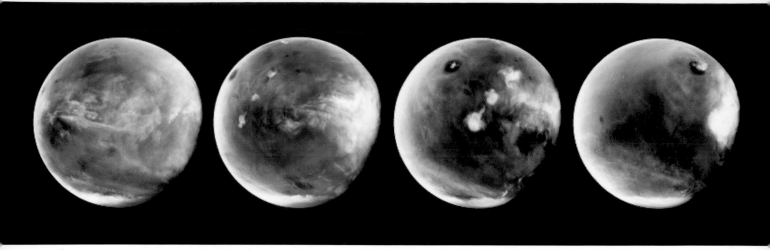

**ABOVE Imaging ultraviolet spectrograph observations of clouds by MAVEN in July 2016.** *(NASA/Maven/ Univ. of Colorado)*

The results indicated that the deterioration of the atmosphere is significantly increased during solar storms, when the planet is blasted by the plasma of the solar wind. That loss of atmosphere to space probably played a key role in the gradual shift (believed to have occurred between 4.2 and 3.7 billion years ago) from a dense carbon dioxide-dominated atmosphere that would have kept the planet sufficiently warm for liquid water to be stable on the surface, to the cold and arid environment we see today.

This transition was made possible by the deterioration of the global magnetic field as the cooling of the planet's core shut down the dynamo that produced the protective magnetosphere. Molecules of gases in the upper atmosphere are now dissociated by solar ultraviolet and energetic charged particles of the solar wind, and the ions are 'picked up' and carried away into space.

**RIGHT This oblique perspective of the Argyre basin by a Viking orbiter also shows the band of the planet's atmosphere.** *(NASA/JPL-Caltech)*

# Sniffing for methane

To follow up on the detection by terrestrial telescopes and by Mars Express of methane in the atmosphere of Mars, the European Space Agency's Trace Gas Orbiter entered a steeply inclined orbit around the planet in October 2016.

At the time of writing this book, the spacecraft was still executing a lengthy aerobraking campaign. However, during periapsis passes it was able to test its instruments to confirm that they were working satisfactorily.

Once it has achieved a circular orbit, it is to scan the atmosphere to create profiles from the surface up to 160km to characterise both spatial and temporal variation of methane and other trace gases, and in particular attempt to localise their sources. If the methane is found in the presence of propane or ethane this will strongly imply a biological origin, but if it is accompanied by gases such as sulphur dioxide this will mean it is a by-product of geological activity. Either way, extant life or active geology, the result will be significant.

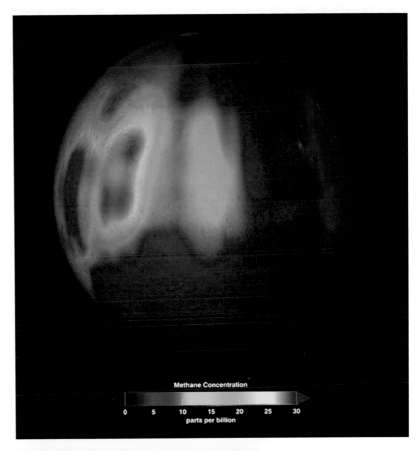

Methane Concentration

| | | | | | | |
|---|---|---|---|---|---|---|
| 0 | 5 | 10 | 15 | 20 | 25 | 30 |

parts per billion

**ABOVE** In 2003 terrestrial telescopes detected methane in the atmosphere of Mars, where it was northern summer. Surprisingly, it was not evenly distributed but varied regionally and temporally. *(NASA/GSFC/Trent Schindler)*

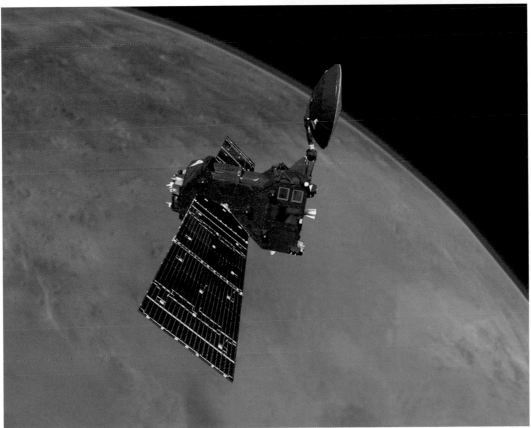

**LEFT** The Trace Gas Orbiter spacecraft. *(ESA)*

*Chapter Six*

# Seeking ground truth

Robotic rovers serving as field geologists have revolutionised the study of Mars. They have been sent to investigate sites where orbital evidence suggested water had played a part early in the history of the planet. In addition, landers have been sent to undertake specific measurements at fixed sites. It is doubtful anyone prior to the Space Age could have imagined this activity.

**OPPOSITE A skycrane preparing to deliver the Curiosity rover to the surface of Mars.** *(NASA/JPL-Caltech)*

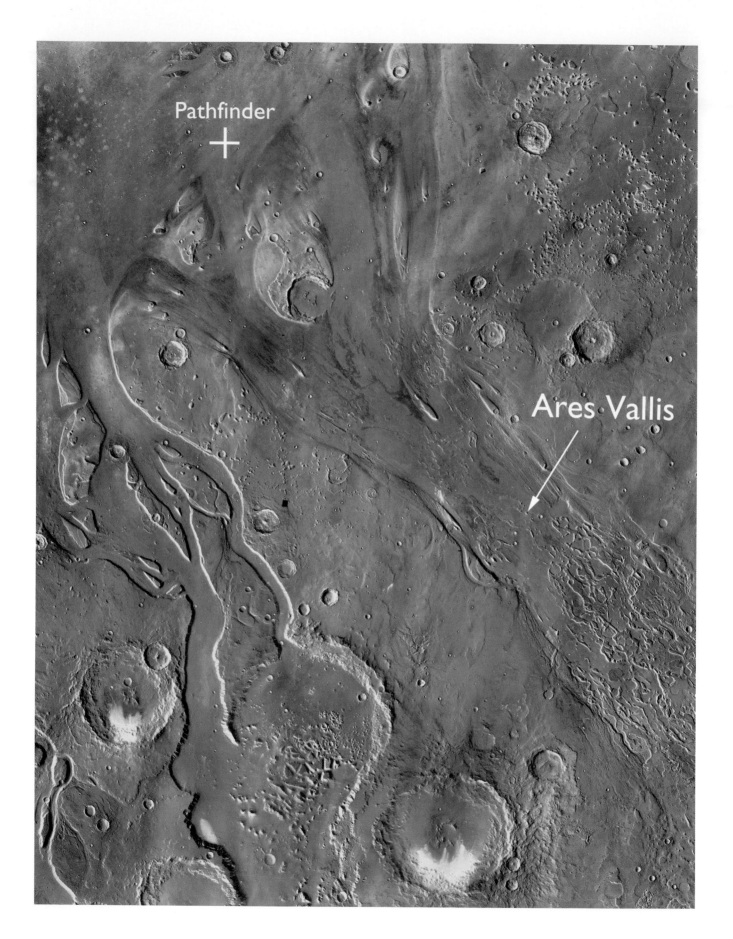

Pathfinder

Ares Vallis

# Pathfinder

**N**ASA established the Discovery Program to demonstrate the practicality of low-cost deep space missions.

The first one, Mars Pathfinder, was launched during the 1996 window and it reached its target on 4 July 1997.

Unlike the Vikings there was no orbiter; the lander entered the atmosphere from the approach trajectory. The primary engineering objective was to test an entry, descent and landing system capable of delivering a small lander to a site that was considerably rougher than could have been assigned to a Viking-style mission.

In turn, it employed a heat shield, a parachute, rocket thrusters, and a set of airbags. A radar altimeter inflated the airbags at a height of 300m, the thrusters fired at 50m in order to slow the rate of descent, and the package was released with the still-firing thrusters drawing the parachute away. The package bounced and rolled across the surface until coming to a halt. Shortly thereafter, the bags deflated, the lander unfolded a set of petals, and the bags were retracted out of the way.

In addition to requiring to be at a low elevation to maximise the potential for parachute braking, the target had to lie in the equatorial zone in order to power the solar panels.

A site was selected where the Ares Vallis outflow channel debouched onto Chryse Planitia. This was rough terrain about 850km southeast of the Viking 1 site, close to the target originally selected for that lander. Given the geology, it was hoped to find rocks which originated from sources all the way back to the southern highlands.

The Pathfinder lander was equipped with a stereoscopic camera and a suite of meteorology sensors, but its primary payload was Sojourner, an experimental six-wheeled rover. As the light-travel time meant the rover could not be controlled from Earth in real

**OPPOSITE** The Ares Vallis outflow channel drains from the southern highlands onto the low-lying Chryse Planitia. The Mars Pathfinder landing site is marked on this image, taken by the Mars Odyssey orbiter. *(NASA/JPL-Caltech/Arizona State Univ.)*

## THE PATHFINDER LANDER

The Viking landers had touched down softly on Mars employing rocket thrusters. Pathfinder was to demonstrate that an airbag system could deliver a package to the surface at speeds that would otherwise be damaging.

The tetrahedron-shaped lander had a triangular base and three side petals that were folded for transit. After the parachute phase of the descent, the airbags were to be rapidly inflated by gas generators. Each petal had an airbag made up of six 1.8m lobes configured in a 'billiard rack' pattern. Tests during development established that the airbags could withstand striking a rock as tall as 0.5m with a vertical velocity of 14m/sec and a horizontal velocity of 20m/sec.

After landing, the base petal deployed an imager consisting of 23mm f/10 optics and a pair of 256 x 256-pixel CCDs, each having a wheel with a dozen filters. This was on a head that could rotate in azimuth and elevation, and was to deploy on a spring-coiled telescoping mast to a height of 1.5m to provide a reasonable point of view of the landscape and the rover. One of the side petals bore the Sojourner rover.

**BELOW Inspecting the multiple-lobed airbags developed for Mars Pathfinder.** *(NASA/ILC)*

**ABOVE** A panoramic view by the Mars Pathfinder lander across Ares Vallis to two hills that were promptly named Twin Peaks after a TV show. The rock-strewn landscape is a series of ridges and swales. The Sojourner rover has driven down the ramp onto the surface and is inspecting a rock that is much larger than itself, nicknamed Yogi. *(NASA/JPL-Caltech)*

**RIGHT** Because the Sojourner rover was so small, its perspective made the landscape appear very dramatic. *(NASA/JPL-Caltech)*

pebbles lying around that may have been shed in this manner, and their rounded shape suggested considerable erosion prior to being conglomerated. Also, the fact that the conglomerates were delivered to this site intact implied that they had been formed either in a riverbed or in the aftermath of an earlier flood. Other conglomerates, in particular the large angular fragments, were more likely to be ejecta from the impact that excavated a crater several kilometres to the south.

Sojourner confirmed the finding by the Viking landers that the soil was rich in sulphur, strengthening the theory that global dust storms have homogenised the dust planet-wide. There was a highly magnetic mineral present in the soil which, on Earth, is formed when an iron-rich aqueous solution is freeze-dried. This lent support to the idea that Mars was once considerably warmer and wetter than it is now.

In addition to the expected rocks of low-silica basalt, there was evidence that some have a high silica content. The latter was a surprise because it suggested the planet was more thermally evolved than believed, and that in turn raised the issue of when the internal 'heat engine' had shut down.

Later it was realised the rocks were coated with a rind of weathered material that wasn't chemically representative of the underlying minerals. The engineers belatedly wished they

had fitted the rover with a tool to scrape a rock clean prior to using the spectrometer.

It was widely agreed that the landing site was an ancient flood plain but this did not in itself prove Mars had once possessed an active hydrological cycle, as an intense but brief eruption of water from the ground could have scoured deep channels and transported boulders even in a cold and dry climate. To prove the planet had been warm and wet over a geologically significant interval, scientists would have to locate and study outcrops of sedimentary rock.

## Frustrations

When Mars Express approached Mars in December 2003, it released a landing probe supplied by a British team. In homage to HMS *Beagle*, the ship on which Charles Darwin made his epic voyage in the 1830s, the lander had been named Beagle 2.

It would be the first mission since the Viking landers explicitly to seek life on Mars. Its task was to inspect rocks for minerals that would show the presence of liquid water in the past, seek carbonaceous structures left by organisms living in that water, and measure the ratio of the carbon isotopes as a 'biomarker' test.

As regards the possibility of there still being life, Beagle 2 was not to make a Viking-style test designed to prompt metabolic activity by micro-organisms in the soil. Instead, it was to analyse the atmosphere for out-of-equilibrium gases that could be the result of metabolism by, for example, methanogenic archea. Unlike the Vikings, which analysed the topsoil which was irradiated by solar ultraviolet, solar wind, and cosmic rays, it was to investigate more benign environments by extracting cores from rocks and by using a 'mole' on a cable that would burrow into the soil to a depth where conditions would not only be less hostile than the surface but might also be moist.

It was also to measure the ambient ultraviolet insolation, the rate of oxidation by the atmosphere, and the oxidation state of the iron in the soil in order to test the post-Viking hypothesis that the ultraviolet flux was producing superoxides in the atmosphere that were accumulating in the soil at a concentration sufficient to break down organic molecules.

The landing site had to be low-lying for maximum parachute braking and in the equatorial zone for solar power. Isidis Planitia on the line of dichotomy was chosen as a large sediment-filled basin with what seemed to be a fairly smooth surface.

The vehicle entered the atmosphere on 19 December as planned. Following the initial aerodynamic deceleration it was to deploy a parachute and then inflate airbags to cushion the

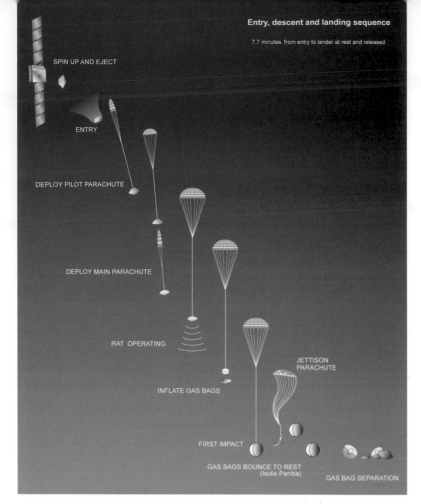

**Entry, descent and landing sequence**

7.7 minutes from entry to lander at rest and released

SPIN UP AND EJECT

ENTRY

DEPLOY PILOT PARACHUTE

DEPLOY MAIN PARACHUTE

RAT OPERATING

INFLATE GAS BAGS

JETTISON PARACHUTE

FIRST IMPACT

GAS BAGS BOUNCE TO REST
(Isidis Panitia)

GAS BAG SEPARATION

**LEFT The entry, descent and landing sequence for the Beagle 2 mission.** *(ESA/OU)*

impact. The signal to indicate that it had landed was never received. As it was not equipped to transmit in-flight telemetry, there was little for the investigators to go on. All manner of possible failure modes were considered without reaching a definitive conclusion.

However, in January 2015 it was located intact on the surface by the super-high-resolution camera on Mars Reconnaissance Orbiter. Evidently the landing went to plan but when the craft attempted to deploy its various parts, two of its four solar panels failed to unfold and this blocked the communications antenna.

We shall have to wait for another mission to directly test for microbial life in the subsoil.

Meanwhile, entirely independently of the Discovery Program, the new Mars Exploration Program called for sending landers to carry out in-situ observations designed to answer specific scientific questions.

In the Martian spring the south polar cap consisting of carbon dioxide frost retreats to expose a 'layered terrain' that is characterised by alternating bands which represent different mixes of dust and water ice. The idea was that these would shed light on how the climate had changed during recent times, in much the same way as can be done on Earth using 'tree rings'. The first mission was targeted in this region.

The site had to be as far north as possible in order to provide solar power, so a 'tongue' of the layered terrain that projected 15° (800km) out from the pole was chosen which would have been exposed only a few weeks prior to the arrival of the lander.

The development of Mars Polar Lander

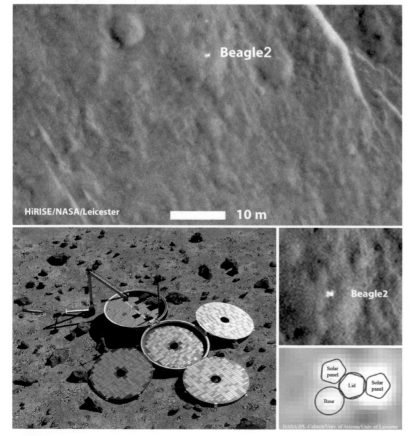

Beagle2

HiRISE/NASA/Leicester          10 m

Beagle2

Solar panel

Lid

Solar panel

Base

NASA/JPL-Caltech/Univ. of Arizona/Univ. of Leicester

**LEFT There was no signal to indicate that Beagle 2 had landed, so it was presumed to have crashed into the surface. It was later identified in imagery provided by Mars Reconnaissance Orbiter. Although it had landed successfully on Isidis Planitia, not all of its solar panels had unfolded, which had prevented it from communicating.** *(NASA/JPL-Caltech/Univ. of Arizona/Univ. of Leicester)*

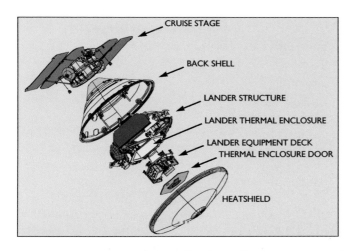

CRUISE STAGE

BACK SHELL

LANDER STRUCTURE

LANDER THERMAL ENCLOSURE

LANDER EQUIPMENT DECK
THERMAL ENCLOSURE DOOR

HEATSHIELD

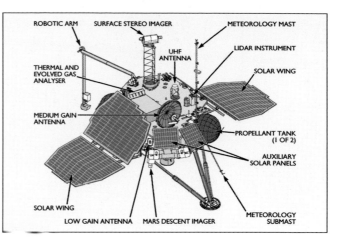

ROBOTIC ARM   SURFACE STEREO IMAGER   METEOROLOGY MAST

UHF ANTENNA

LIDAR INSTRUMENT

THERMAL AND EVOLVED GAS ANALYSER

SOLAR WING

MEDIUM GAIN ANTENNA

PROPELLANT TANK (1 OF 2)

AUXILIARY SOLAR PANELS

SOLAR WING

LOW GAIN ANTENNA   MARS DESCENT IMAGER

METEOROLOGY SUBMAST

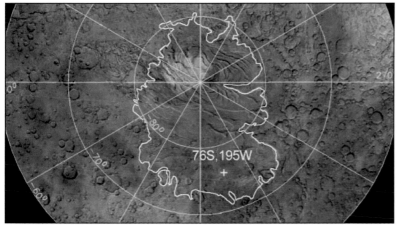

76S, 195W

**ABOVE LEFT** Details of the Mars Polar Lander entry system. *(NASA/JPL-Caltech/Woods)*

**ABOVE** Details of Mars Polar Lander. *(NASA/JPL-Caltech/Woods)*

**LEFT** The target for Mars Polar Lander was just within the boundary of the south polar cap at a point where the carbon dioxide seasonal cap would just have receded. *(NASA)*

**BELOW** The entry, descent and landing for Mars Polar Lander. *(NASA/JPL-Caltech/Woods)*

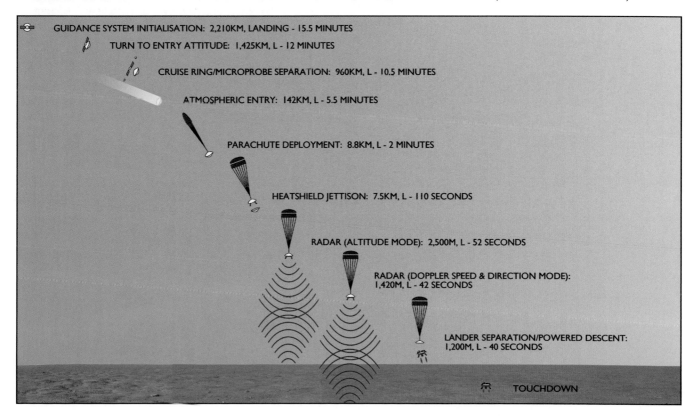

GUIDANCE SYSTEM INITIALISATION: 2,210KM, LANDING - 15.5 MINUTES

TURN TO ENTRY ATTITUDE: 1,425KM, L - 12 MINUTES

CRUISE RING/MICROPROBE SEPARATION: 960KM, L - 10.5 MINUTES

ATMOSPHERIC ENTRY: 142KM, L - 5.5 MINUTES

PARACHUTE DEPLOYMENT: 8.8KM, L - 2 MINUTES

HEATSHIELD JETTISON: 7.5KM, L - 110 SECONDS

RADAR (ALTITUDE MODE): 2,500M, L - 52 SECONDS

RADAR (DOPPLER SPEED & DIRECTION MODE): 1,420M, L - 42 SECONDS

LANDER SEPARATION/POWERED DESCENT: 1,200M, L - 40 SECONDS

TOUCHDOWN

started before Mars Pathfinder had demonstrated the airbag system so the entry, descent and landing method was similar to Viking, except that, like Pathfinder, the probe entered the atmosphere directly from the approach trajectory.

Its arrival on 3 December 1999 proved to be another disaster for NASA and the Mars scientists. The vehicle was not to transmit telemetry during its descent, so when it failed to report in later the investigators had very little evidence to sift. Nevertheless, after several months of analysis it was decided that the parachute had deployed properly at an altitude of 7km and then 10sec later the heat shield was released and the lander's three legs were deployed. The radar acquired the surface at an altitude of 1,500m. With 40sec to go, the backshell was jettisoned, the lander ignited its engines, cancelled its horizontal drift and began

to regulate its sink rate. To that point, things had gone to plan.

At a height of 40m, having attained the required rate of descent, the computer started to monitor the signal that would indicate one of the legs had contacted the surface.

It was here that the investigators uncovered a design flaw that ought to have been caught in testing during development. This flaw had caused the vital signal to be asserted by the process of deploying the legs, so the computer terminated the engine burn immediately. Although the drop to the ground would have taken just a few seconds, the vehicle must have been wrecked by the shock of impact.

Due to the time required to redesign the lander to prevent all of the possible failure modes that could have afflicted Mars Polar Lander, and to prove its worth by exhaustive testing, it wasn't feasible to launch a lander in the 2001 window to repeat the lost mission.

## Robotic field geology

In fact, NASA opted to postpone the polar investigation in order to introduce an advanced type of rover, one with far greater mobility than Sojourner and a more sophisticated set of sensors. And for redundancy it was decided to send two of them.

By this point, the major question was whether the early climate of Mars had been warm and wet, and the strategy to investigate this was to 'follow the water'.

The plan was to select sites on the basis of the orbital data from Mars Global Surveyor and Mars Odyssey and seek ground truth to test hypotheses about the early history of the planet.

As robotic field geologists, the Mars Exploration Rovers were to be equipped to investigate the physical characteristics, chemistry and mineralogy of individual rocks in order to determine whether they were either formed in or were altered by liquid water.

The engineering requirements for the landing sites were that they be at low elevation to provide the greatest parachute braking, be in the equatorial zone to optimally illuminate the solar panels, and not be so rocky as to inhibit

**BELOW Preparing a Mars Exploration Rover. The vehicle was much larger than its Sojourner predecessor.** *(NASA/JPL-Caltech/KSC)*

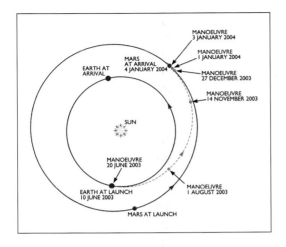

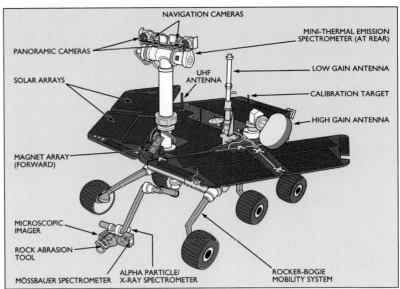

NAVIGATION CAMERAS

MINI-THERMAL EMISSION SPECTROMETER (AT REAR)

PANORAMIC CAMERAS

LOW GAIN ANTENNA

SOLAR ARRAYS

UHF ANTENNA

CALIBRATION TARGET

HIGH GAIN ANTENNA

MAGNET ARRAY (FORWARD)

MICROSCOPIC IMAGER

ROCK ABRASION TOOL

MÖSSBAUER SPECTROMETER

ALPHA PARTICLE/ X-RAY SPECTROMETER

ROCKER-BOGIE MOBILITY SYSTEM

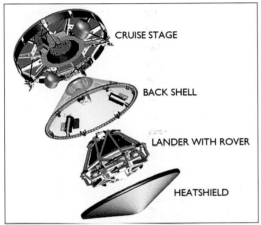

CRUISE STAGE

BACK SHELL

LANDER WITH ROVER

HEATSHIELD

**ABOVE LEFT** The interplanetary trajectory of the Mars Exploration Rover named Spirit. *(NASA/JPL-Caltech/Woods)*

**ABOVE** Details of a Mars Exploration Rover. *(NASA/JPL-Caltech/Woods)*

**LEFT** Details of the Mars Exploration Rover entry system. *(NASA/JPL-Caltech/Woods)*

**BELOW** The entry, descent and landing for a Mars Exploration Rover. *(NASA/JPL-Caltech/Woods)*

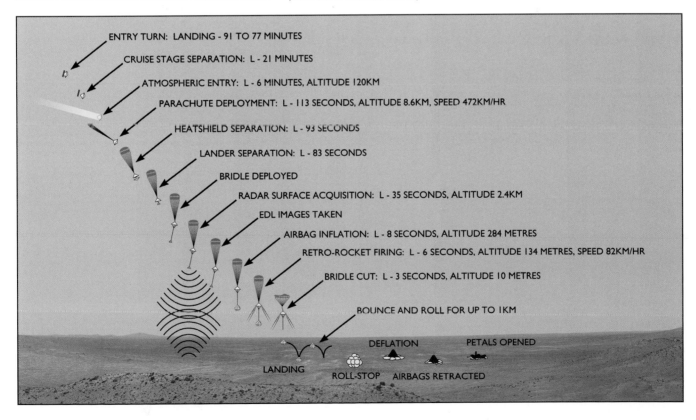

ENTRY TURN: LANDING - 91 TO 77 MINUTES

CRUISE STAGE SEPARATION: L - 21 MINUTES

ATMOSPHERIC ENTRY: L - 6 MINUTES, ALTITUDE 120KM

PARACHUTE DEPLOYMENT: L - 113 SECONDS, ALTITUDE 8.6KM, SPEED 472KM/HR

HEATSHIELD SEPARATION: L - 93 SECONDS

LANDER SEPARATION: L - 83 SECONDS

BRIDLE DEPLOYED

RADAR SURFACE ACQUISITION: L - 35 SECONDS, ALTITUDE 2.4KM

EDL IMAGES TAKEN

AIRBAG INFLATION: L - 8 SECONDS, ALTITUDE 284 METRES

RETRO-ROCKET FIRING: L - 6 SECONDS, ALTITUDE 134 METRES, SPEED 82KM/HR

BRIDLE CUT: L - 3 SECONDS, ALTITUDE 10 METRES

BOUNCE AND ROLL FOR UP TO 1KM

DEFLATION

PETALS OPENED

LANDING

ROLL-STOP

AIRBAGS RETRACTED

The first MER target was Gusev, a 150km crater located just south of the line of dichotomy. The morphology strongly suggested that the southern rim had been breached by Ma'adim Vallis, a channel that ran for 900km and originated in the highlands, forming a lake that had then left a smooth deposit of sediment in the crater.

The rover, named Spirit, was to land on this putative lakebed to search for evidence of water action. In particular, scientists wanted to distinguish between sediments formed on the bed of a body of standing water and rocks which were formed in running water, and also any rocks that formed in the absence of water and were later modified by its presence. If, however, the rover found the floor of the crater littered primarily with volcanic rocks, this would imply the lakebed (if it had actually existed) was buried by a blanket of volcanic material, possibly from Apollinaris Patera to the north.

The site for the second rover, named Opportunity, was selected on the basis of data from the thermal emission spectrometer carried by Mars Global Surveyor. This showed a concentration of a mineral called grey hematite in an exceptionally flat patch of Meridiani Planum just south of the line of dichotomy that overlies the cratered Arabia Terra and has the appearance of a mantle of pyroclastics several hundred metres thick.

The reddish hue of the planet derives from the iron at its surface having been oxidised. Hematite comes in red and grey forms that are chemically the same but differ in the size of their crystals. The fine-grained red hematite that eroded from exposed rock has been distributed planet-wide by global dust storms and hence was of no relevance to the search for evidence of past water.

To accumulate crystals of hematite into the large grains of grey hematite on Earth usually (but not always) requires the presence of liquid water. Opportunity was to investigate whether the grey hematite at Meridiani Planum was present in layers of sediment in a lake, in veins resulting from the alteration of pre-existing rocks by a hydrothermal system, or perhaps the product of another process that didn't involve water.

**ABOVE** The morphology suggests that when Ma'adim Vallis breached the southern wall of the large crater Gusev this produced a lake in the crater. This image is a mosaic from the Viking orbiters. *(NASA/JPL-Caltech/USGS)*

**BELOW** The dark streaks on the floor of the crater Gusev were produced by dust devils. The image combines data from Mars Express and Mars Odyssey. The cross indicates where Spirit landed and the Columbia Hills complex lies to the southeast. *(NASA/JPL-Caltech/ESA/Arizona State Univ.)*

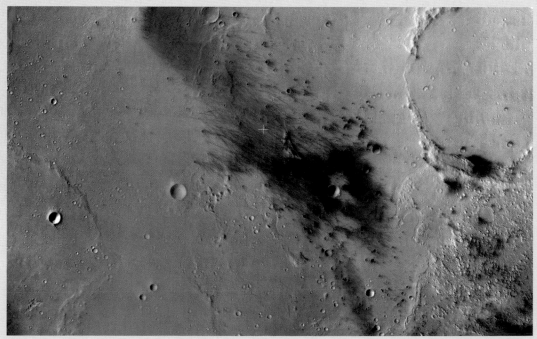

mobility. It was decided to select one site for its morphological characteristics and the other on the basis of chemical remote-sensing from orbit.

## Spirit in Gusev crater

There was concern as Spirit approached Mars on 3 January 2004, as infrared data from Mars Global Surveyor indicated that a regional dust storm on the far side of the planet had warmed the upper atmosphere on a global basis. It was calculated that the diminished air density would reduce the efficacy of the first phase of the braking procedure. Since a deceleration sensor was to initiate the deployment of the parachute, the delayed action would cut the interval between deploying the chute and the radar firing the retrorockets by about 20%, to only 90sec. That event sequence would nominally require 100sec, so the computer was told to deploy the chute several seconds early. This was a calculated risk, because if the dynamic pressure was too great it would rip the canopy.

After the loss of Mars Polar Lander, NASA had stipulated that future landers must indicate their progress. Hence a low-gain antenna on the entry system was to transmit a series of 10sec tones at specific points in the descent sequence.

Travelling at 5.4km/sec, the lander penetrated the atmosphere at an altitude of 125km. The deceleration peaked at 6g and the temperature of the heat shield peaked at 1,500°C. On slowing to 400m/sec the parachute deployed at a height of 7km. The engineers cheered when the deployment tone was received but the most intensive portion of the sequence, involving dozens of pyrotechnic devices, was still to come.

Once the forward heat shield had been jettisoned, the backshell unreeled a 20m tether supporting the lander. At this point, the lander started to transmit its own UHF telemetry to Mars Global Surveyor, which recorded the data for later relay to Earth.

The radar altimeter was activated at an altitude of 2.4km and a downward-looking

## MER CAPABILITIES

Whereas Sojourner was very low to the ground, the cameras for MER would be atop a mast to provide a viewpoint similar to that of a human explorer.

The panoramic stereoscopic camera provided a resolution matching 20/20 eyesight and had 13 filters in the visible and near-infrared spectrum in order to analyse the surface chemistry.

The new rover was to be able to operate semi-autonomously to a greater extent than was possible for Sojourner, and a monochrome navigation camera would record a wide horizon arc to enable the operators on Earth to monitor its motions and sampling activities.

The third instrument to exploit the mast was a smaller version of the thermal emission spectrometer on Mars Global Surveyor. Its electronics were housed in the body of the rover and the mast served as a periscope. Its sensor produced a mosaic of 'false colour' circular spots that were later superimposed on an image. Each spot was a spectrum in 167 wavelengths and the longer the sensor 'stared' at a target the better was its signal-to-noise ratio. It was to assist in the selection of rocks for individual sampling using the tools on a short robotic arm which was capable of motions involving five degrees of freedom.

The arm had a microscopic imager with a resolution comparable to that of a hand lens. It was to assist in identifying rocks that were formed in water, features of volcanic and impact origin, and veins of minerals left by the presence of water in a rock. For soils it would show the sizes and shapes of the grains, and provide insight into erosional processes.

Like Sojourner, the new rover had an alpha particle X-ray spectrometer that could detect all the main rock-forming elements apart from hydrogen. Since iron interacts strongly with liquid water, there was also a Mössbauer spectrometer to investigate iron-bearing minerals.

Because the rocks investigated by Sojourner had been coated with a rind of weathered material that impaired the spectrometer studies, the new rover had a brush of stainless steel bristles to sweep dust from a rock and a rotary grinder to expose a circle several millimetres deep to enable the spectrometer to analyse the rock rather than any superficial layer.

The Spirit landing site in Gusev proved to be much flatter and less rocky than the Viking and Mars Pathfinder sites. The first impression was that it resembled the ancient lakebed that was the rationale for its selection. *(NASA-JPL-Caltech/Cornell)*

camera snapped pictures at altitudes of 2, 1.75 and 1.5km. These were analysed on board to estimate the horizontal motion due to near-surface winds.

At a height of 284m the airbags were inflated. There were six bags affixed to each of the four faces of the lander, each a double-layered bladder to resist their puncture by rocks. At 100m, side rockets on the backshell fired to cancel out the horizontal drift calculated from the three descent images, then a trio of powerful retrorockets on the backshell were ignited to cancel the rate of descent. With the lander halted at a height of 9m, the tether was cut and the still-firing retrorockets drew the backshell and parachute clear.

The tone to indicate contact with the surface prompted another tremendous cheer from the engineers at JPL.

The airbags caused the lander to bounce several times before coming to a halt. Working autonomously it started to deflate and retract the airbags, first the base petal upon which the rover was installed and then the three side petals. As the lander could have come to rest in any orientation the powered hinges on the side petals were capable of flipping the lander onto its base. In fact, it had come to rest the right way up.

At the landing site it was early in the afternoon on what would be defined as sol 0. As soon as the petals had deployed, the rover unfolded its solar panels to recharge its batteries.

When Mars Odyssey flew overhead several hours later the lander uplinked data. This included the first monochrome imagery. It showed that there were no rocks to block the path of the rover when it later drove off the lander. As further images streamed in they showed a landscape that was much flatter than earlier landing sites, which was consistent with a lakebed. There were a lot of rocks of various sizes, but nothing that would prevent Spirit from driving around.

After shutting down for the night Spirit awakened 2hr after sunrise, used its navcam to locate the Sun in the sky, calculated the offset for Earth, elevated the high-gain antenna, and slewed it around to establish direct communications. If it had failed to attain this link-up, it would have pursued a predefined sequence of activities for the full primary mission of 90 sols.

Spirit elevated its mast and used the pancam to take the first of a series of high-resolution colour images to scan the horizon. These were sent via satellite relays. Operating at its maximum UHF rate the rover could uplink 50Mbit during each relay pass. Since this bandwidth couldn't be solely devoted to the pancam, transmitting the panorama and the matching mini-TES data required most of the 10 sols or so that had been allocated to the preparations for driving the rover off its lander.

The results showed that this site was remarkably different from those of the Vikings and Pathfinder. There were not as many large boulders. The surfaces of the rocks were very smooth, and their shapes ranged from very rounded to quite angular. This suggested they were very hard, made of fine-grained material, and had been polished by aeolian forces over a protracted period.

In effect, this initial imaging was what an immobile lander would have done. As such, it formed a major objective of the science programme.

What had looked plausibly like a lakebed in the low-resolution monochrome navcam imagery seemed less so in the pancam panorama. A terrestrial lakebed is typically flat and made of very finely grained sediments. The ground at Gusev was littered with rocks, many of which were fractured and looked like they could well be debris thrown in from elsewhere, either by volcanism or by impacts, so if this was a lakebed it was not in pristine condition.

Fortunately, the issue would not have to be resolved by looking at imagery from a stationary lander. The mobility of the rover and its suite of sensors would enable it to act in much the same way that a field geologist would, were we able to send humans to the planet.

Clearly visible in the panorama was the rim of the largest of several craters seen in the descent imagery. It was 200m in diameter and 275m away from the lander. By the manner in which an impact makes a crater, the material from the deepest part of the excavation lies on the rim, which in this case rose 4m above its surroundings. If the lakebed had been buried then the best chance of quickly sampling the sedimentary rock beneath might be in the ejecta excavated by this crater.

On the horizon some 2.3km away to the southeast were a group of hills that rose about 100m above the plain. If these had been submerged by the lake, the fact that they rose above whatever had buried the lakebed suggested they might hold the key to the mission.

The nominal plan for the 90-sol primary mission was to undertake a series of sampling activities ranging out 600m from the lander, but even before Spirit was ready to drive off its lander the science team were selecting targets much farther afield.

It was decided that after preliminary operations in the vicinity of the lander, Spirit would drive to the large crater whose ejecta would provide a 'window' into the subsurface. Any layering in the interior wall might reveal the thickness of the material masking the putative lakebed.

In the longer term, the intention was to head for the hills. On the geological map made from orbital imagery these were 'etched' terrain. They seemed to be part of a stratigraphic unit that was otherwise buried. The closer Spirit was able to approach, the greater were the chances of encountering material on the plain that originated from the hills. And if it operated for long enough, Spirit might be able to cross the 'contact' on the map and undertake an in-situ examination of the lower slopes.

Spirit drove off its lander on sol 12 and carried out some engineering tests. The first sampling task was to examine the soil. The microscope showed this to be a conglomeration of dust particles. After the spectrometers had been tested, the rover moved off to study its first rock.

The team had selected a rock several metres away, named Adirondack. It possessed a flat

face that would readily facilitate grinding to scrape off any rind. After a major computer problem was resolved, the data from the spectrometers was retrieved and the presence of olivine, pyroxene, and magnetite indicated a basaltic chemistry. As the brush was applied it exposed a darker material. This was a surprise, since the rock had been chosen for seeming

BELOW Spirit cleans, drills and investigates the composition of a rock named Adirondack, finding it to be basalt. (NASA/ JPL-Caltech/Cornell/ Univ. of Mainz)

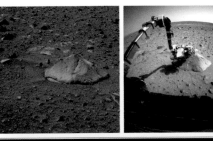

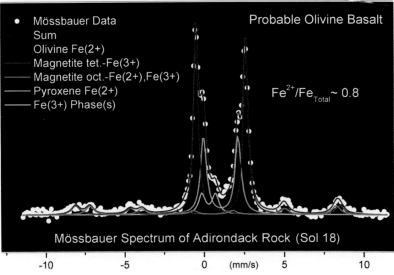

• Mössbauer Data
Sum
Olivine Fe(2+)
Magnetite tet.-Fe(3+)
Magnetite oct.-Fe(2+),Fe(3+)
Pyroxene Fe(2+)
Fe(3+) Phase(s)

Probable Olivine Basalt

$Fe^{2+}/Fe_{Total} \sim 0.8$

Mössbauer Spectrum of Adirondack Rock (Sol 18)

−10      −5      0    (mm/s)    5      10

RIGHT In conjunction
with the thermal
emission spectrometer
aboard Mars Global
Surveyor, a similar,
but miniaturised
instrument on the
Spirit rover derived
a profile of the air
temperature above
Gusev. From its
vantage point, Spirit
could measure only
to an altitude of
about 6km. It was
the first time that a
temperature profile
was measured all the
way from the top of
the atmosphere to
ground level.
(NASA/JPL-Caltech/
GSFC/Arizona State
Univ./Cornell)

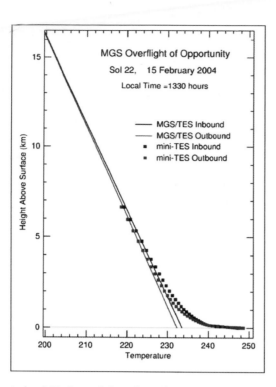

**MGS Overflight of Opportunity**
Sol 22,   15 February 2004
Local Time = 1330 hours

— MGS/TES Inbound
— MGS/TES Outbound
■ mini-TES Inbound
■ mini-TES Outbound

to be fairly free of dust. It verified the Sojourner
lesson that it was essential to clean a rock prior
to measuring its chemistry and mineralogy.
The grinder drilled a hole 2.7mm deep and
the brush swept the dust from the hole. Then
the microscope took a look at the cavity and

the spectrometers analysed the exposure. The
results confirmed the initial assessment; it was a
volcanic rock.

After further sampling close to its lander, Spirit
set off for the large crater that had been named
Bonneville, located 275m to the northeast.

After studying several targets of opportunity
in transit, Spirit encountered a two-fold
increase in the coverage of rocks and a
five-fold increase in their average size as it
penetrated the ejecta blanket of the crater
and climbed the 15° slope onto its rim, where
the navcam took a 180° look at the interior.
There was no evidence of layering in the wall.
The rover slowly drove around the rim, with its
spectrometers analysing the soil.

There was a 2m block on the rim that
must have been dug from the deepest
excavation. Its multifaceted scalloped
surface indicated that it had been scoured
by windborne sand for a very long time. Due
to its light tone and sugary texture, the rock
was examined in detail to determine whether
it represented the putative sedimentary
substrate of the basaltic plain. It was basalt,
but analysis of multiple coatings on the rock,
as well as material in fractures, showed that

minerals had precipitated after water had penetrated into fractures in the rock. But this hadn't been in a lake the alteration had occurred underground.

The fact that the ejecta was basaltic indicated that if Gusev was a lakebed, the sediments must be buried deeper than was excavated by the impact which made Bonneville crater.

By the time that Spirit left Bonneville to head for the hills, it was nearing the end of the nominal 90-sol duration of its primary mission.

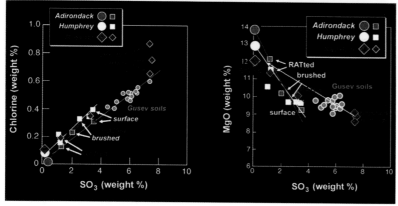

transition from the material representing the plain to that of the hills.

When the 'contact' was encountered on sol 156 it proved to be extremely obvious, with the chemistry changing substantially over a span of only several metres.

As Spirit explored the Columbia Hills, sometimes struggling against wheel slippage, it became clear that the rocks were old and weathered. It discovered ample evidence of rocks that had been altered by sulphur, chlorine, bromine, and potassium, all of which are readily transported by water. Some rocks were more than 50% composed of various salts. Their abundance implied that the rocks on the hill had been infused with water. Once on the hills, the rover did not find any unaltered volcanic rocks at all. And when it was in a valley beyond Husband Hill its wheels exposed a light-toned material which the spectrometers found to be almost pure, finely grained hydrated silica, the formation of which indicated hot water such as a hydrothermal environment.

Spirit was immobilised by soft soil on sol 1,892, but it continued to provide useful data in situ until it fell silent on sol 2,210.

A roving vehicle gave a glimpse of the 'big picture' that would have eluded a static lander such as a Viking.

The morphology of Gusev that was visible in orbital imagery strongly implied it once held a

**ABOVE** On leaving Bonneville, Spirit examined a rock called Mazatzal. After brushing a flower-shaped area sufficient to allow the mast-mounted spectrometer a clear field of view, it drilled a hole to investigate the composition of the interior. *(NASA/JPL-Caltech/Cornell/Max Planck Institute)*

The hills had been named in honour of the astronauts who perished when Space Shuttle Columbia disintegrated during atmospheric re-entry in 2003. The plan was to drive toward the West Spur of Husband Hill, which was the tallest member of the group, to seek the

**RIGHT** The Columbia Hills viewed by Spirit while still on its lander. The seven peaks were named in honour of the astronauts lost aboard Space Shuttle Columbia. *(NASA/JPL-Caltech/Cornell)*

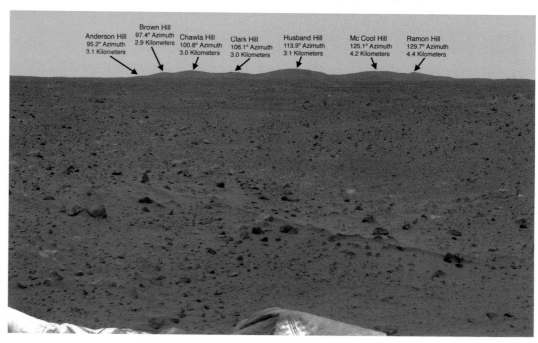

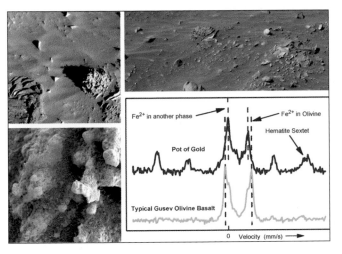

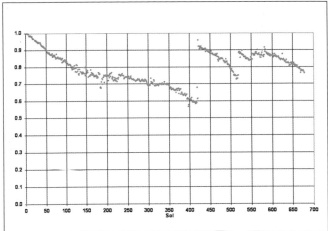

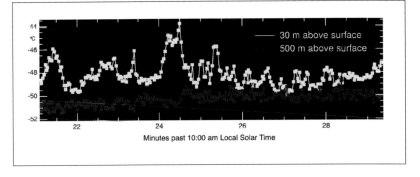

**ABOVE** On reaching the base of the Columbia Hills, Spirit investigated a rock named Pot of Gold. A magnifying lens showed it to be heavily pitted and to possess nuggets on stalks. The analysis showed it to contain hematite. This suggested the material on the hills was different from that of the plain. *(NASA/JPL-Caltech/Cornell/Max Planck Institute)*

**ABOVE RIGHT** It had been expected that the operating lives of the Spirit rover (and its companion Opportunity) would be limited by the accumulation of dust on the solar panels, but dust devils swept the panels clean! *(NASA/JPL-Caltech)*

**ABOVE** Spirit's thermal emission spectrometer found the air 30m above the surface to be somewhat hotter than that at 500m and to fluctuate in temperature to a greater extent. Such information helped scientists to understand how the layer of air closest to the surface interacted with global winds. *(NASA/JPL-Caltech/Cornell/Arizona State Univ.)*

substantial lake. But instead of finding the floor of the crater to be a sedimentary carbonate or evaporite, Spirit found only volcanics, implying that the plain was an ancient lava flow. Although some of the rocks had been chemically altered by water, this did not require them to have been immersed in a lake; the aqueous fluid could have been present as volatiles in the lava or have percolated through the rocks while underground. The rocks on the plain therefore provided no evidence to support the lake hypothesis. And that is where the results would have stood if Spirit had been limited to its nominal 90-sol mission.

**RIGHT** As Spirit ascended the West Spur, it recorded a dust devil crossing the plain below. *(NASA/JPL-Caltech/Texas A&M Univ.)*

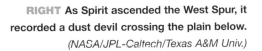

**119**

**ABOVE Digitally superimposing Spirit at the summit of Husband Hill gives a sense of perspective.**
*(NASA/JPL-Solar System Visualisation Team)*

Meanwhile, the dilemma concerning the absence of carbonate, which must surely have settled as sediment on the floor of a lake, was being resolved on the opposite side of the planet by Spirit's twin.

## Opportunity at Meridiani

O n 25 January 2004 the second MER mission performed atmospheric entry in darkness far above Valles Marineris. After being released, the lander bounced and rolled for 20min before finally coming to rest. Although it didn't settle on its base, it easily righted itself.

When Mars Odyssey made the first UHF relay pass 4hr later it forwarded to Earth 20Mbit of data which included dozens of images that portrayed a bizarre, alien landscape. The

Fortunately, Spirit was able to reach the Columbia Hills several kilometres from the landing site. The fact that the peaks rise 100m above the plain did not rule against their being of sedimentary origin, as the floor of Gusev lies 2,500m below its rim. Although Spirit's exploration of the Columbia Hills established the material to be of volcanogenic origin, the extent to which the rocks were altered by aqueous fluid held out the prospect that the hills were once submerged by a lake.

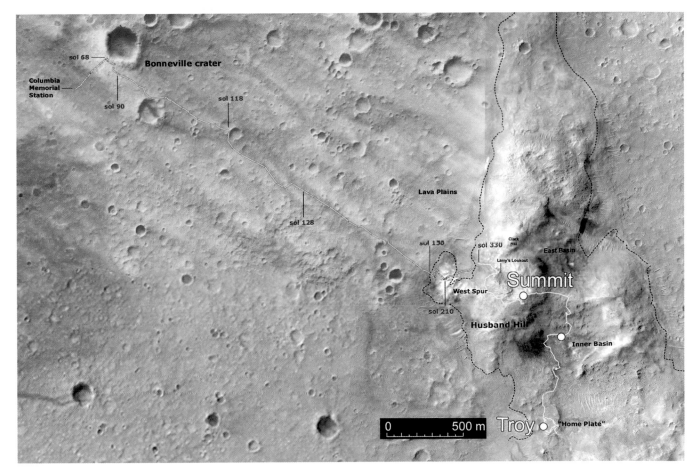

proximity of the horizon showed the lander was sitting in a crater 22m in diameter and 2m deep. The soil seemed to have the consistency of talc and it had perfectly preserved the imprints of the airbags. But what amazed the science team was a horizontal line of light-toned rock in the wall on one side of the crater. Because this appeared to be an exposure of bedrock, the mission had hit the scientific jackpot.

As Stephen Squyres, the scientific principal investigator for the Mars Exploration Rover project has explained, "If you're golfing on a par three hole, an Eagle is a hole in one." So the crater was named Eagle.

Having recharged its batteries, Opportunity went to sleep at sunset. It awoke during the

**ABOVE The overall traverse of the Spirit rover.** *(Adapted from NASA/JPL-Caltech/ Cornell/UA/NMMNHS)*

**BELOW The second MER mission came to rest in a shallow crater that was named Eagle. The Opportunity rover could not peer over the rim to the Meridiani plain beyond, but geologists were delighted to see an outcrop of layered rock in the wall.** *(NASA/JPL-Caltech/Cornell)*

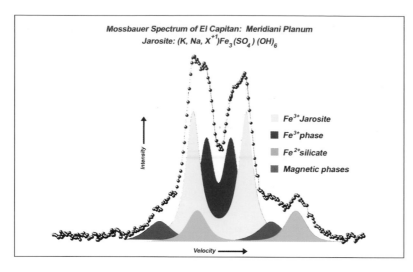

Mossbauer Spectrum of El Capitan: Meridiani Planum
Jarosite: $(K, Na, X^{+1})Fe_3(SO_4)(OH)_6$

- Fe³⁺Jarosite
- Fe³⁺phase
- Fe²⁺silicate
- Magnetic phases

*Intensity*

*Velocity →*

**ABOVE** The yellow peaks in this analysis of the El Capitan outcrop indicate the presence of jarosite, a mineral that contains water in the form of hydroxyl as a part of its structure. This was 'smoking gun' evidence for water-driven processes. *(NASA/JPL-Caltech/Univ. of Mainz)*

**RIGHT** On finding a cluster of blueberries in a shallow cavity on the outcrop of Eagle crater, Opportunity analysed both a cluster of berries and a patch of adjacent rock that was free of them. Although the spectrum for a group of berries included typical outcrop characteristics, it also exhibited an intense hematite signature (labelled 'magnetic sextet'). Meridiani Planum was selected as the target for the second MER mission because orbital sensors had detected a high concentration of hematite, which is an iron-bearing mineral often formed in water. *(NASA/JPL-Caltech/Cornell/Univ. of Mainz)*

Sol 46  Bare outcrop
Sol 48  Berry bowl

Mössbauer spectra

Hematite sextet

*Intensity →*

0

*Velocity (mm/sec) →*

lander had set down in this region, far from any craters, it would have found the site littered with hematite spherules and yet, lacking any sense of the underlying bedrock, would not have been able to provide any insight into their origin.

The next task was obvious. Opportunity must drive to a much larger crater in order to inspect its walls for insight into the deeper

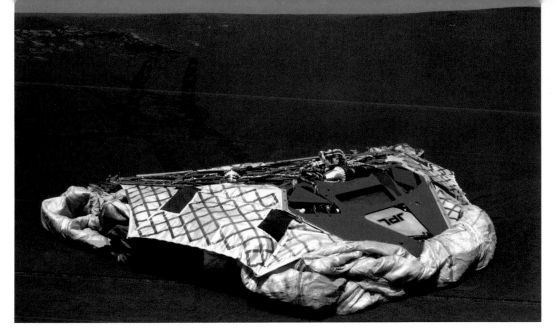

structure of the plain. By good fortune there was such a crater named Endurance, 750m to the east.

The open plain was remarkably devoid of rocks, but there was one close by Eagle crater. Because its chemistry differed from the local bedrock, it must have been tossed in from somewhere else.

Some sinuous features 100m into the journey that were barely resolvable in orbital imagery proved to be a chain of deep 'fractures' that were edged by rocks similar to the outcrop in Eagle crater. Opportunity navigated a safe route around this obstacle.

Next was a crater some 8m in diameter that was 450m from the landing site. It possessed an ejecta blanket and a blocky interior. If Opportunity had landed in this crater, the task of interpreting the rock would have been rather more difficult than in Eagle crater.

By the end of sol 93, Opportunity was 70m from the near rim of Endurance and its view of the upper portion of the far rim stirred considerable excitement in the science team because there was a large mass of dark rock.

The rover slowed as it made its approach, ascended a shallow incline, and crept to within 50cm of the lip of the rim, beyond which the surface fell away at 18° to the floor at a depth of 20m. It was now possible to see beneath the rock unit that was outcropped in Eagle crater. There were several exposures with a considerable amount of layering.

The planners knew that to properly investigate the layering the rover would need to drill holes in it, but there was a distinct possibility that if it entered the crater it might not be able to escape.

After surveying the crater from a number of points on its rim, the engineers decided that the safest route was from the southwest.

Opportunity positioned itself on the lip, then slowly nudged its front wheels over the crest, then all six wheels, then reversed out to evaluate the traction on the blueberry-strewn 18° incline. Finally, it descended several metres

down the slope which exposed at least three layers showing distinct contacts. And there were many more ahead. By investigating the layers to identify the stratigraphic column the rover would be doing exactly what a field geologist would do in that situation.

Even after progressing 7m down the slope at the entry point the rover was still sampling sulphates. This showed that the influence of water was present to a considerable depth. The fact that the darker sulphates lacked evidence of water ripples implied that after having been laid down in water the material had been homogenised, probably by the wind stirring it up at times when the surface was dry. The clue was the ratio of chlorine to bromine. At Eagle this ratio varied on a scale of tens of centimetres. Elements having different solubilities will precipitate out under different conditions. Thus variability in the ratio was

clear evidence of an evaporative process. If the outcrop at Eagle were to be homogenised, then it would gain a uniform ratio as observed at Endurance. So there had been times when the surface was wet and times when it was dry. On Earth this landform is known as a playa.

Some rocks farther into the crater gave the appearance of being coated with dried mud, hinting at the influence of water both prior to and after the crater was formed, as if water had pooled in the cavity.

And when it was deep inside Endurance, the rover was able to look up at the large exposure named the Burns Cliff on the southern wall. It was very significant because the outcrop extended far below the layers now known to be composed of sulphates.

Scientists were particularly eager to inspect where two of the layers in the cliff intersected at an angle. This huge crossbed suggested wind

**RIGHT Opportunity looks up at the Burns Cliff from a vantage point on the inner slope of Endurance crater.** *(NASA/JPL-Caltech/Cornell)*

action, with the beds being laid down at the observed angle. On Earth this would be clear evidence of a petrified dune. Confirmation that the lowest layers of the cliff had been transported by wind meant that prior to the episodic shallow surface water that left sediments, Meridiani had been a desert of basaltic sand dunes.

Having spent six months exploring Endurance, Opportunity emerged from the crater on sol 315, where it paused to recharge its battery.

A few sols later, it drove a short distance to a rock which the mast-mounted spectrometer showed to be rich in iron and nickel. In fact, it was without doubt a meteorite.

Since there is reason to believe Mars ought to have been hit by many more rocky meteorites than iron ones, this discovery prompted the suggested that some of the small rocks on the planet might be rocky meteorites.

**ABOVE** After emerging from Endurance crater, Opportunity went to inspect the heat shield that it had discarded during its descent. Alongside, it found an iron meteorite that was promptly named Heat Shield Rock (top). Later, it found another meteorite called Block Island (bottom). *(NASA/JPL-Caltech/Cornell)*

**LEFT** After departing from Endurance crater, Opportunity turned south for Victoria crater. Along the way, however, it became stuck in a ripple of sand which was soon labelled the Purgatory Dune. With the rover stuck, the upper image looks back along its line of approach. It took fully a month to escape and the lower image shows that withdrawal. *(NASA/JPL-Caltech/Cornell)*

**BELOW** On the way to Victoria, patches of the underlying bedrock began to appear between the dunes. To reduce the risk of once again being trapped in soft material, as much as possible the rover drove along the 'highways', even though in some places these were littered with cobbles. *(NASA/JPL-Caltech/Cornell)*

**ABOVE** Continuing towards Victoria, Opportunity saw streaky high-altitude clouds. If the first space mission had landed at this place, the view would have reinforced the early impression of the planet as an arid desert! *(NASA/JPL-Caltech/Cornell)*

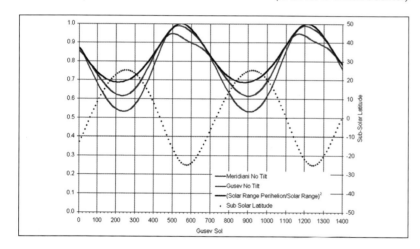

**LEFT** This chart illustrates the variation in available solar power for Spirit and Opportunity over approx. two Martian years. Two factors affect the available power: the eccentricity of the planet's orbit and the tilt of its axis. The horizontal scale is the sols since Spirit's arrival. The vertical scale on the left shows the available solar power as a ratio of the amount available at the equator when Mars is nearest to the Sun. The red line indicates power availability at Spirit's landing site (Gusev). The blue line indicates power availability at Opportunity's landing site (Meridiani). The vertical scale on the right applies to the dotted line, and is the latitude north or south of the planet's equator where the noon Sun is overhead at different times of the local year. *(NASA/JPL-Caltech)*

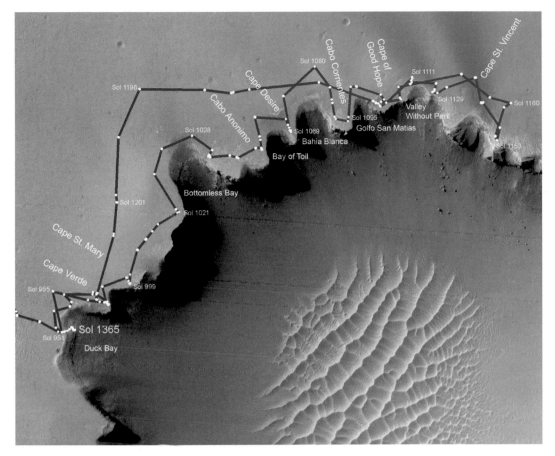

They would be hard to spot at any other
location on the planet, but on the bland plain
they drew attention to themselves.

Having provided insight into the bedrock
and past history of Meridiani Planum by
its explorations of Eagle and Endurance,
Opportunity set off south on sol 358, despite
being well past its nominal mission in terms of
duration and distance.

Driving long distances on successive sols,
Opportunity inspected a number of craters of
various sizes on its way to the etched terrain,
where the light-toned rock was exposed at the
surface in between pronounced sand ripples
that occasionally brought the rover to a halt.

To minimise the risk of the rover becoming
trapped in a dune, it was directed to drive
on 'highways' in between dunes and to jump

**LEFT** A false-colour view of the rock exposure of Cape St. Vincent, one of the promontories on the rim of Victoria. The material at the top of the stack consists of loose, jumbled rock. A bit farther down there is an abrupt transition to solid bedrock, this being marked by a bright band of rock that is visible around the entire crater. The bright band seems to represent the surface immediately before the impact that excavated the crater. *(NASA/JPL-Caltech/Cornell)*

**ABOVE** After Victoria there was a long drive southeast for the even larger crater Endeavour. Opportunity entered this hilly terrain in approaching Cape York on the rim of the crater. *(NASA/JPL-Caltech/Cornell/Arizona State Univ.)*

dunes at safe-looking points. In this manner, it zigzagged southeast towards Victoria, a crater 800m in diameter whose very jagged rim looked like it must expose a lot more stratigraphy than was visible in the Burns Cliff, and indeed it did.

A risk analysis indicated Opportunity could safely descend the shallowest of the slopes. It started at Duck Bay to examine a succession of three 'bathtub ring' layers in the wall. These

**LEFT** On the rim of Endeavour, Opportunity entered Marathon Valley and ascended Knudsen Ridge. Looking back, it saw a dust devil on the plain. *(NASA/JPL/Don Davis)*

**RIGHT** Opportunity's traverse from the landing site to Endeavour crater with major science stops at Endurance and Victoria craters. (NASA/JPL-Caltech/MSSS/NMMNHS)

had different textures and grain sizes but were similar in composition.

The top layer was fine- to medium-grained and finely layered with abundant blueberries. The middle layer was smoother and lighter-toned, and the layering was even finer. The lower layer was darker, probably due to containing almost-black basaltic sand. The sulphur-bearing compounds in all cases implied that the water must have been of a very high salinity. No known terrestrial lifeform could survive in such a fluid.

Continuing downslope, Opportunity found the final exposed outcrop at Duck Bay to be richer in iron in the form of hematite than any Mars rock yet measured.

It was therefore concluded that the minerals in the lowest layers of Victoria had formed in less acidic conditions. Possibly the water had become ever more acidic as volcanism injected sulphur into the atmosphere. If life had developed on the planet, the increasing acidity of the water may well have killed it off.

After spending 340 sols inside Victoria negotiating sandy and steep slopes, Opportunity made its exit.

Although the venerable rover had already achieved so much, the JPL team opted to attempt an excursion of 12km farther to the southeast with the objective of reaching Endeavour, a crater 22km in diameter whose 300m depth offered an even greater insight into the history of Meridiani.

Orbital imagery showed layering in the wall of Endeavour that infrared data suggested contained iron-bearing and magnesium-bearing clays created by the interaction of olivine with water. Clays would have formed in conditions far less acidic than those that produced sulfates and therefore less hostile to life. When the rover finally arrived it confirmed the orbital data.

In 2013 Opportunity marked ten years on Mars, during which it had driven a distance of 35km. At the time of writing this book in the

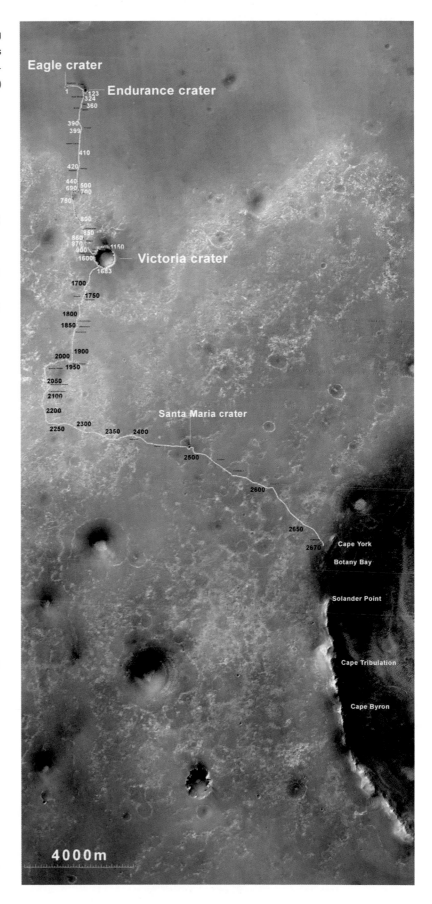

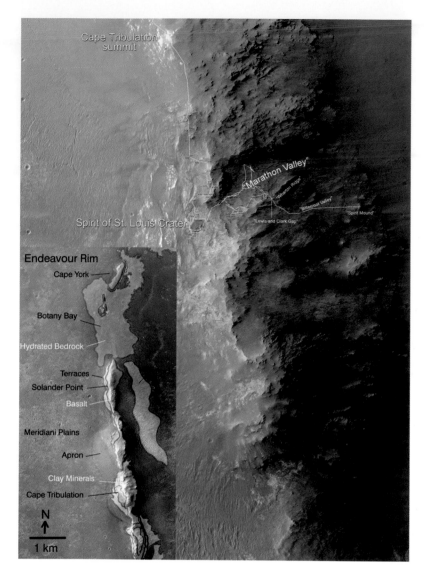

Cape Tribulation summit

Marathon Valley

"Wharton Ridge"

"Bitterroot Valley"

"Lewis and Clark Gap"

"Spirit Mound"

Spirit of St. Louis Crater

**Endeavour Rim**

Cape York

Botany Bay

Hydrated Bedrock

Terraces

Solander Point

Basalt

Meridiani Plains

Apron

Clay Minerals

Cape Tribulation

N

1 km

**LEFT** Detail of Opportunity's exploration of the western rim of Endeavour crater using data from the HiRISE and CRISM instruments of Mars Reconnaissance Orbiter. The route is complete up to the writing of this book in 2017. Note that the rover is still operating at the time of writring. *(NASA/JPL-Caltech/JHUAPL)*

summer of 2017 it is still exploring the western rim of Endeavour!

By any measure, the scientific results of the Spirit and Opportunity rovers more than justified their cost.

# Phoenix finds ice

With renewed confidence NASA opted to try again with Mars Polar Lander, now renamed Mars Phoenix, using improved versions of some of the instruments lost in 1999 and others from the cancelled 2001 mission. It would continue the 'follow the water' strategy of the Mars Exploration Program.

However, the target was switched from the south polar region to the seasonal carbon dioxide cap in the north, where Mars Odyssey had detected the presence of near-surface ice.

It was to investigate the history of water and water ice at the landing site, how the climate is affected by the dynamics of the atmosphere above the polar region, and whether the polar regions could support life.

The target was on Vastitas Borealis, about 1,200km from the north pole and 4.1km below the 'sea level' datum. Orbital imagery indicated it to be a level plain free of rocks. Its distinguishing characteristic was a pattern of polygonal 'pillows' several metres wide that were presumed to derive from the seasonal expansion and contraction of ice, as occurs in a terrestrial permafrost zone.

Phoenix arrived on 25 May 2008. The telemetry marking its progress was recorded by Mars Odyssey. Mars Reconnaissance Orbiter was able to take a picture of it descending by parachute. At an altitude of 960m the backshell was jettisoned. The lander fell freely for a few seconds then ignited its engines and touched down safely. For the first time there were three active missions on the surface of Mars.

**BELOW** Assembly and test of the Mars Phoenix lander at the manufacturer. *(NASA/JPL-Caltech/UA/Lockheed Martin)*

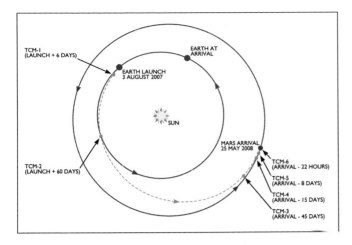

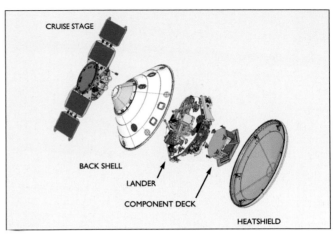

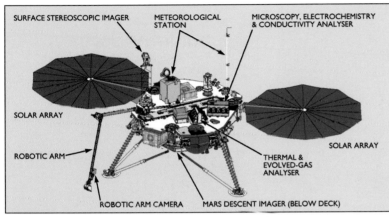

**ABOVE LEFT** The interplanetary trajectory of the Phoenix mission. *(NASA/Woods)*

**ABOVE** Details of the Phoenix entry system. *(NASA/Woods)*

**LEFT** Details of the Phoenix lander. *(NASA/Woods)*

**BELOW** The entry, descent and landing procedure for the Phoenix lander. *(NASA/Woods)*

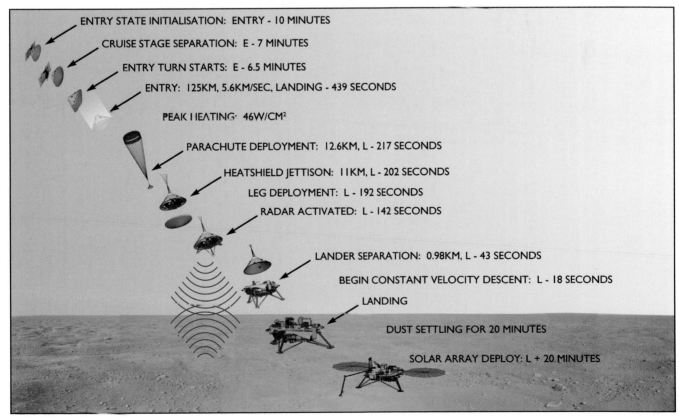

ENTRY STATE INITIALISATION: ENTRY - 10 MINUTES

CRUISE STAGE SEPARATION: E - 7 MINUTES

ENTRY TURN STARTS: E - 6.5 MINUTES

ENTRY: 125KM, 5.6KM/SEC, LANDING - 439 SECONDS

PEAK HEATING: 46W/CM²

PARACHUTE DEPLOYMENT: 12.6KM, L - 217 SECONDS

HEATSHIELD JETTISON: 11KM, L - 202 SECONDS

LEG DEPLOYMENT: L - 192 SECONDS

RADAR ACTIVATED: L - 142 SECONDS

LANDER SEPARATION: 0.98KM, L - 43 SECONDS

BEGIN CONSTANT VELOCITY DESCENT: L - 18 SECONDS

LANDING

DUST SETTLING FOR 20 MINUTES

SOLAR ARRAY DEPLOY: L + 20 MINUTES

ABOVE **Material stirred up by the arrival of the Phoenix lander has settled on top of a footpad.** *(NASA/JPL-Caltech/Univ. of Arizona)*

After waiting 15min for the dust stirred up by its arrival to settle, the lander deployed its solar panels, camera, and weather masts. It established an uplink with Mars Odyssey when that passed overhead about 2hr later.

This being the late northern spring, the Sun shone on the solar panels the entire time. By the solstice on 25 June the Sun would reach its maximum elevation of 47.0°. Phoenix would not experience a sunset until September, by which time its primary mission would be complete.

The site was extremely level, and characterised by polygons several metres across and separated by narrow troughs 20 to 50cm deep. There was a litter of small rocks, but larger ones were scarce. There were no dunes or ripples, but a range of low hills poked over the horizon.

Fortuitously, the robotic arm could reach to sample both the frozen ice-rich soil of the polygon on which the vehicle stood and the ice-poor

RIGHT **The Phoenix lander looks out to the horizon on the Vastitas Borealis at 68°N. It shows the polygonal pattern that is so distinctive in orbital imagery and is littered with pebbles, but there are no boulders. The polygonal cracking is believed to have been caused by the seasonal contraction and expansion of surface ice. A similar pattern has been observed in permafrost terrains on Earth.** *(NASA/JPL-Caltech/Univ. of Arizona)*

Sol 20     Sol 24

2/3"

**ABOVE** The Phoenix lander's robotic arm at work, with a sample in its scoop. *(NASA/JPL-Caltech/Univ. of Arizona/Texas A&M Univ.)*

**RIGHT** Upon repeatedly inspecting its trench, Phoenix saw particles of ice sublimating to vapour. *(NASA/JPL-Caltech/Univ. of Arizona/Texas A&M Univ.)*

soil in one of the troughs. Upon inspecting the ground immediately beneath itself, Phoenix found the soil had been eroded by the braking rockets, exposing a bright hard-looking feature that could have been rock but looked remarkably like a slab of ice. This suggested the landscape was a blanket of soil several centimetres thick over an ice table. It was an excellent start to the scientific mission. Temperatures ranged from –80°C in the early morning to a relatively mild –30°C in the afternoon. This and the pressure of 0.086 millibars verified that the ice was water, since

carbon dioxide would have rapidly sublimated at such temperatures.

Although small pieces of ice in the soil were observed to sublimate, the ice table itself remained stable. It was rock-hard and not even the tungsten carbide blade of the scoop could scrape a sample off it.

The main instrument was the thermal and evolved gas analyser inherited from Mars Polar Lander. This had miniature ovens to process samples of soil provided by a robotic arm. The instrument was first to apply a mild warming

**LEFT** When the robotic arm peered beneath the Phoenix lander it showed that the landing jets had blown away the thin mantle of fine material and exposed a slab of ice below. This discovery led one member of the science team to exclaim "Holy Cow!" *(NASA/JPL-Caltech/Univ. of Arizona/ Texas A&M Univ.)*

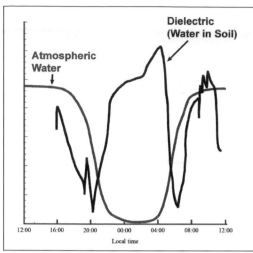

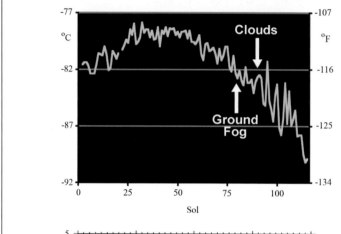

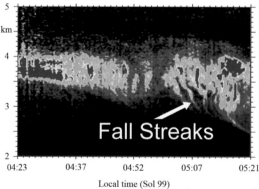

**Fall Streaks**

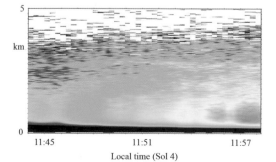

Local time (Sol 4)

**ABOVE LEFT** When the arm of the Phoenix lander delivered a soil sample to the thermal and evolved gas analyser, the material was so cohesive it resisted passing through the filters. *(NASA/JPL-Caltech/Univ. of Arizona/Max Planck Institute)*

**ABOVE** A simplified presentation of data from overnight measurements by the thermal and electrical conductivity probe of the Phoenix lander. Water in the soil appears to increase overnight, just as water in the atmosphere disappears. *(NASA/JPL-Caltech/Univ. of Arizona)*

**LEFT** Phoenix provided atmospheric data. As the minimum temperature increased through the summer season, the atmospheric humidity also increased. Clouds, ground fog, and frost were observed each night after the temperature started dropping. In addition, the lidar beamed at the zenith measured clouds and airborne dust. In the centre panel, the vertical streaks at the base of the cloud on the right of the image show ice crystals falling from the cloud, similar to snow. The streaks are curved as the winds are faster around 3km than at higher altitudes. Scientists inferred that the snow was water-based rather than carbon-dioxide snow because Martian temperatures are currently too warm for the latter. The bottom panel shows lidar data for a 15min period approaching noon on sol 4. Towards the end of the measurement (on the right-hand side), higher concentrations of dust (depicted in red and orange) pass over the lander. *(NASA/JPL-Caltech/Univ. of Arizona/ Canadian Space Agency)* at 35°C to melt ice, then raise the temperature to 175°C to release gases, before finally baking what remained at 1,000°C to break up organics, salts, and water-altered minerals. It was to profile the absorbed heat as a function of the applied temperature, because the temperature at which any water was released would assist in identifying the kinds of hydrated minerals present.

Unfortunately, although a thermal and electrical conductivity probe reported the soil near the surface to be mostly dry, it was sufficiently cohesive to become stuck in the sieves that fed the ovens, not passing through until after the ice in it had sublimated in the polar summer sunshine, and this tended to undermine an experiment designed to study the history of water. But this, as one scientist put it, was "largely because Mars failed to cooperate".

If it could have measured profiles of salt distribution with depth, this might have revealed much about the ebb and flow of water during any brief spells of warmer climate. Evidently an investigation of this type will require some sort of drill or subsurface probe.

A significant result was provided by the wet chemistry experiment. This had four cells with pure water. After a soil sample had been soaked, it was tested for salts, acidity, alkalinity, and oxidation potential in order to determine whether the ice present beneath the surface would be conducive to life.

A sample taken from several centimetres depth near the centre of a polygon showed a slightly alkaline pH of about 8.3. The Viking landers had suggested an acidic soil rich in a strongly oxidising compound that was tentatively identified as hydrogen peroxide. The addition of acid to the solution ought to have lowered its pH, but it proved to remain almost stable, as if something was 'buffering' it, such as some form of carbonate. In fact, this characteristic of carbonates would make the Martian surface rather less inimical to life. Salts included magnesium, sodium, potassium, and calcium. Also chlorine salts, but few sulphates and no nitrates. The most interesting result and possibly the most important of the entire mission, was a fairly high concentration of ions of perchloric acid salts. This was interesting for the reason that certain types of bacteria

## THE PH SCALE

In chemistry, pH (potential of hydrogen) is a numeric scale used to specify the acidity or alkalinity of an aqueous solution. It is approximately the negative of the base 10 logarithm of the molar concentration of hydrogen ions in units of moles per litre.

Pure water is neutral at pH 7; solutions with a pH less than 7 are acidic and those with a pH greater than 7 are alkali.

and plants on Earth use perchlorate salts as a supply of energy.

Despite the pessimistic conclusions drawn from the Viking results, the soil at the high-latitude Phoenix site was "surprisingly hospitable to life", being no more hostile than the 'dry valleys' of Antarctica. However, Phoenix didn't have a mass spectrometer to test for the presence of organic materials.

Phoenix made one of the most visually striking observations toward the end of its mission, namely an early morning snowfall extending down to the surface.

The mission was declared complete in August, with winter looming and the return of the seasonal polar cap. Later orbital imagery showed that the weight of ice had caused one of its solar panels to collapse.

## Curiosity in Gale crater

As it stood in 2000, the Mars Exploration Program envisaged launching a heavy rover in 2007 to demonstrate the technologies that would be needed by a sample-return mission, particularly the ability to make a precision landing. Its target ellipse was to be at least an order of magnitude smaller than those of the Pathfinder and Mars Exploration Rover missions. If the test succeeded, NASA hoped to progress to a sample-return mission early in the next decade.

But a change in policy from the White House prompted a revision of funding priorities. Mars missions scheduled within the decade remained in place, but the sample-return was pushed downstream.

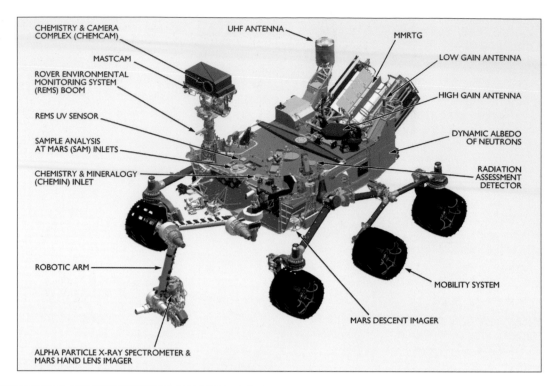

**RIGHT** Details of the Curiosity rover built for the Mars Science Laboratory mission. *(NASA/JPL-Caltech/ Woods)*

CHEMISTRY & CAMERA COMPLEX (CHEMCAM)

MASTCAM

ROVER ENVIRONMENTAL MONITORING SYSTEM (REMS) BOOM

REMS UV SENSOR

SAMPLE ANALYSIS AT MARS (SAM) INLETS

CHEMISTRY & MINERALOGY (CHEMIN) INLET

ROBOTIC ARM

ALPHA PARTICLE X-RAY SPECTROMETER & MARS HAND LENS IMAGER

UHF ANTENNA

MMRTG

LOW GAIN ANTENNA

HIGH GAIN ANTENNA

DYNAMIC ALBEDO OF NEUTRONS

RADIATION ASSESSMENT DETECTOR

MOBILITY SYSTEM

MARS DESCENT IMAGER

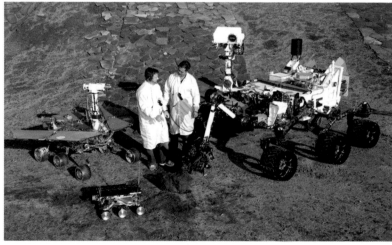

**LEFT** Matt Robinson (left) and Wesley Kuykendall in the Mars Yard at JPL along with three generations of Mars rover. In front is the flight spare for the first Mars rover, Sojourner, delivered by Mars Pathfinder. On the left is an MER test rover identical to Spirit and Opportunity. And on the right is a test version of the MSL Curiosity rover. *(NASA/JPL-Caltech)*

Meanwhile, the objective of the new rover was changed from a technology demonstration to a proper scientific mission called the Mars Science Laboratory.

Although the launch was postponed initially to 2009 and then to 2011, this allowed time to switch the mode of landing to a 'skycrane', a development of the scenario where an airbag-lander suspended on a tether was brought to a hover by firing rockets in the backshell and then released to fall a short distance to the ground, bounce, and roll to a stop.

The skycrane, officially named the Powered Descent Vehicle, would hover several metres above the surface, reel the rover down to the

**LEFT** Preparing the Curiosity rover. *(NASA/JPL-Caltech)*

Map labels:
- Phoenix
- Viking 2
- Viking 1
- Mars Pathfinder
- Opportunity
- Gale
- Spirit
- Holden & Eberswalde

ground, release it, and then fly clear. This would enable heavier payloads to be delivered to Mars and be emplaced much more gently than was possible using airbags.

At 900kg, the big new rover, named Curiosity, was almost twice the mass of Spirit or Opportunity combined with its landing stage. Those missions were right on the operating limit of the airbag system.

Without the benefit of a precursor demonstration, the challenge would be to make a skycrane work first time. It would be a risky venture. If it were to fail, the telemetry would be essential for figuring out the problem. As the projected costs increased, the option of launching a pair for redundancy was ruled out.

Like the Viking landers, Curiosity got its electrical power from a radioisotope thermoelectric generator.

Free of solar power restrictions, the planners were able to consider landing sites anywhere within 60° of the equator and as high as 1km above 'sea level'. The ability to use aerodynamic lift to steer during atmospheric entry meant that the target ellipse could be reduced to several tens of kilometres in diameter. And since the skycrane could accommodate slopes steeper than the airbag system it was possible to consider rougher sites.

Landing site workshops started in 2006 and no fewer than 35 targets were initially offered. In fact, this highlighted the existence of a 'rover gap'. In view of discoveries by Mars Express and Mars Reconnaissance Orbiter, there were so many interesting sites that it was not possible to make follow-up studies due to the paucity of surface missions in the budget.

Observations by the Mars Reconnaissance Orbiter spectrometer drew the selectors to sites where clays were present, as these materials could preserve traces of ancient life, but the geology of some of the richest sites was not clear. This tipped the balance in favour of sites showing hydrated minerals where the geology was better understood.

But this emphasis on geology had its weakness, in that Spirit was sent to Gusev because its floor was widely believed to be a lake bed but proved to be mostly covered with basalt.

There were three leading candidates.

One was Gale, a crater 154km in diameter that was located on the line of dichotomy. It had a central peak 5km tall whose layering recorded how surface water had become increasingly acidic over time. It had been considered for a Mars Exploration Rover, but the 100km landing ellipse could not have fitted in without the inclusion of dangerous terrain features.

**ABOVE** The final candidate landing targets for the Mars Science Laboratory mission. As a clay-bearing site where a river once flowed into a lake, Eberswalde crater might provide concentrations of carbon chemistry crucial to the development of life. The peak inside Gale crater might provide a route to drive up a 5km sequence of layers, to investigate a transition from environments that produced clay deposits near the bottom to later environments that produced sulphate deposits. And in Holden crater, running water had created sediments as alluvial fans and catastrophic flood deposits.
*(NASA/JPL-Caltech)*

**LEFT** Data from the thermal emission imaging system of Mars Odyssey superimposed on an orbital image of Gale crater. The false colours signify mineralogical differences in surface materials. For example, windblown dust appears pale pink and olivine-rich basalt is purple. The bright pink on the floor of the crater appears to be a mixture of basaltic sand and windblown dust. The blue near the summit of the central peak probably comes from local materials that are exposed there. The typical average Martian surface soil looks greyish-green.

*(NASA/JPL-Caltech/Arizona State Univ.)*

**BELOW** The interplanetary trajectory, entry system, and entry, descent and landing procedure for the MSL mission. While the vehicle was on its parachute it was imaged by Mars Reconnaissance Orbiter, with the resulting image later being colourised.

*(NASA/JPL-Caltech/Univ. of Arizona/Woods)*

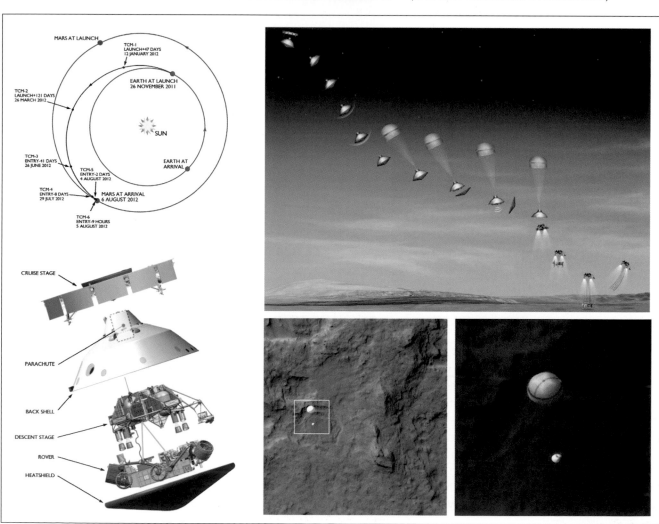

Another excellent candidate was Holden, a 140km crater in the southern highlands. Its wall had been breached by a channel whose outflow deposited layers on the floor. Nearby was the 67km crater Eberswalde. This had a well-preserved delta where an ancient river discharged into a lake and formed clay sediments.

By 2011 the selectors decided that they could not convincingly prove that Holden had held a lake, as there was no evidence of a river discharging into it. And in order to resolve a deadlock between Eberswalde and Gale the decision was referred up to managers and mission scientists, who opted for Gale.

The 21 x 14km ellipse was positioned north of the central peak, near what appeared to be a fan of debris that had spilled off the northern wall.

After investigating the landing site, Curiosity was to navigate a dune field to reach the base of the central peak. Then a valley would provide direct access to layers, the lowest of which were hematite and iron-rich clays that might provide mineralogical and morphological clues of the past existence of water. Geologists wanted to find out whether this material was sediment deposited in a lake or the accumulation of wind-blown material from impacts and volcanic eruptions. Then the rover would ascend the slope to inspect the sulphate layers. Finally, if it lasted long enough, the vehicle might even reach the summit.

On 1 August 2012, as the spacecraft approached Mars, the latest tracking predictions of the entry point were uploaded to its navigation system so that the computer could steer during the hypersonic phase of the atmospheric entry. All three active orbiters – Mars Odyssey, Mars Express and Mars Reconnaissance Orbiter – would be overhead at that time.

On arrival day, 6 August, the cruise stage was jettisoned to burn up in the planet's atmosphere. Ten minutes later the rover, cocooned in the heat shield, entered the atmosphere at an altitude of 125km and travelling at 6.1km/sec. (It would later be calculated that it hit the atmosphere 200m off the aim point, at a speed 11cm/sec off, and with an angle 0.013° shallower than the 15.5° target.) When the speed was Mach 1.7 at an altitude of 10km, the vehicle deployed its parachute.

**LEFT An artist's dramatic depiction of the skycrane lowering the Curiosity rover to the surface.** *(NASA/JPL-Caltech)*

At about this time Earth fell below the local horizon as expected, ending the direct telemetry link of status tones. The remaining phases of the descent were followed using a 'bent pipe' relay via Mars Odyssey.

The radar activated when the forward heat shield was jettisoned. As it had with Phoenix, Mars Reconnaissance Orbiter was able to catch an image of the newcomer on its parachute 1min from landing.

The parachute slowed the speed to 100m/sec. At an altitude of 1.6km the backshell was released, taking the parachute away with it. The skycrane then activated its eight engines. In addition to slowing the rate of descent they nulled the horizontal motion. At a height of 20m the skycrane began to unreel the tether holding the rover.

Events having gone precisely as planned, Curiosity made contact with the surface at a vertical velocity of 0.75m/sec and a horizontal velocity of 4cm/sec. Located just 2.4km east and 400m north of the aim point, this was indeed a 'pinpoint' landing.

Since delivery by skycrane eliminated the need for a landing stage, it was possible to make Curiosity similar in size to a small automobile. The six 50cm wheels were installed on an enlarged version of the rocker-bogie system which had been proven by Sojourner, Spirit, and Opportunity.

Imagery from the navigation camera during the first day (sol 0) showed the central peak of Gale dominating the southern horizon. During planning this was informally referred to as Mount Sharp but it was now officially Aeolis Mons.

A subsequent higher resolution view by the science camera of the hills near the base of the mountain showed a line where the texture changed. This clearly matched the boundary between where orbital sensors had indicated clays to be present and where clays were not. Remarkably, the terrain above the boundary had layers dipping at a steeper angle.

Having scrutinised orbital images, the science team decided to forsake the putative alluvial fan to the north and instead head for a spot some 400m to the east that marked a 'triple contact' by surface units which included layered bedrock, a heavily cratered and possibly old terrain, and the dusty unit that the rover had landed on.

After the software for descent and landing had been replaced with that to drive and to operate the robotic arm and various checks had been undertaken, the rover was ready to start work on sol 15.

In the wake of the ambiguous results of the Viking life-detection tests, the NASA landers had addressed the issue of biology only obliquely in terms of a 'follow the water' strategy.

Now Curiosity would 'follow the carbon' to investigate the possibilities of ancient or extant life. To do this it would catalogue carbon-bearing molecules, hydrogen, nitrogen, oxygen, phosphorus, and sulphur; determine the nature of organic compounds; and identify chemicals such as methane that might be of biological origin. It was capable of detecting as little as 100 parts per trillion of methane, which was far less than was being reported by orbiters and ground-based telescopes. If methane occurred at several tens of parts per billion, the instrument would be able to distinguish the isotopic ratios of carbon within the methane molecules. On Earth this would be a proxy for its organic or chemical origin.

Like Spirit and Opportunity, it would be able to brush, abrade, and drill into rocks to gain a clean surface for spectroscopic analysis. An innovation was an instrument on the mast that could fire a laser at a rock at long range and use a telescopic spectrometer to analyse the results.

Curiosity was also to extend the investigation by the Phoenix lander of the oxidants in the soil far from the polar zone. In addition, a neutron detector could infer the presence of hydrogen, either in the form of ice or hydrated minerals in the near-subsurface. In fact, it would be able to detect as little as 0.1% of water down to a depth of 2m.

Initially the drives were short in order to check out the hazard identification and automatic navigation algorithms, but they became longer as the controllers gained confidence.

On sol 40, Curiosity passed near a rocky outcrop that resembled "a broken sidewalk". It was made of smaller fragments of rock loosely cemented together. The embedded gravel had rounded shapes and comprised large pebbles up to several centimetres in size which seemed to have been transported there by a long-lasting flow of water, perhaps associated with the fan to the north, prior to becoming bound into a sandy matrix. Only remote sensing was carried out, but the laser established the presence of hydrated minerals. The conclusion was that it had been the bottom of a shallow stream.

Reaching the triple contact on sol 56, Curiosity found a drift of fine sand on which a crust had had time to form. The sand was comprised of a lighter toned component that was finely grained and a darker one with larger grains that was similar to terrestrial volcanic soils. When samples were heated progressively to 825°C they liberated significant quantities of water vapour, carbon dioxide, and sulphur dioxide.

It appears some of the carbon dioxide released at higher temperatures was attributable to the decomposition of a carbonate. Sufficient water was issued to perform the first measurement of the deuterium-to-hydrogen ratio on Mars. This revealed that water in the ancient drift was about five times richer in deuterium than terrestrial oceans, which implied that most of the lighter isotope had been lost from the atmosphere over the ages, and probably very early in the planet's history. The hydrogen in the sample indicated that the sample contained either hydrated minerals or molecules of atmospheric water inside the crystal lattices. Such material probably accounted for a large part of the hydrogen detected by instruments on Mars Odyssey and Mars Express.

The triple contact offered a variety of targets for sampling: sandstones with bright quartz-like intrusions, slabs of bedrock, and features similar to terrestrial fossilised mud bubbles.

Moving position, Curiosity entered a low-lying area on sol 125 displaying a succession of layers resembling mudstone and sandstone. A hole 6.4cm deep made by the drill was dull grey, suggesting that the interior of the rock had not oxidised. The powdered rock that travelled out along the flutes of the drill was retained for analysis; it proved to be igneous material with high percentages of clay-like minerals. The levels of suphur-rich and iron-rich clays implied that they were formed in moderately salty water that had a pH that was either neutral or slightly acidic. This was in contrast to the evidence of acidic water Opportunity found at Meridiani Planum. This data implied that where Curiosity sampled was once a lake bed with a watery environment that was present for a considerable time and could have sustained microbial life.

From the isotopic ratios of the noble gases it was possible to calculate how long a rock had been exposed to cosmic rays. This showed that the sediments just sampled had been on the surface for 80 million years. Of course this didn't mean that they formed at that time; they formed early in the planet's history and remained buried until being exhumed by wind action. This 'exposure age' didn't bode well for the search for organics because the bombardment by cosmic rays over such a length of time could easily dissociate all such molecules. Scientists therefore made it a priority to locate a sample of more recently exposed bedrock in which organics might still be preserved.

Having finished its work in the vicinity of the triple contact, Curiosity started the drive south, heading for the base of Aeolis Mons. Along the way, it would examine targets of interest identified in the orbital imagery.

Having initially found no atmospheric molecules that might indicate current methanogenic microbial activity on the planet, measurements taken periodically over 20 months showed increases in late 2013 and early 2014 that averaged 7 parts per billion for methane in the atmosphere. Then there was a decline to one-tenth of that level. Variability was consistent with orbital studies which had shown

the concentration of methane to vary both geographically and over time.

One significant result was the first detection of nitrogen being released after heating surface sediments on the planet Mars. The nitrogen was present in the form of nitric oxide. Because

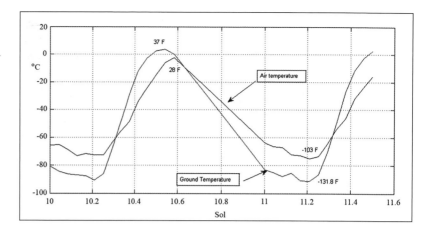

**ABOVE** Air and ground temperatures measured across a single day by the Curiosity rover. As expected, variations in air temperature were less extreme than for the ground temperature. *(NASA/JPL-Caltech/CAB-CSIC-INTA)*

**BELOW** Curiosity measured a spike in the concentration of methane in the atmosphere above Gale crater of up to about 7 parts per billion by volume over a period of several weeks in late 2013 and early 2014, during the first southern hemisphere autumn (northern spring) of its investigations. Although this spike did not repeat during the second Martian year, Curiosity detected a long-term variation in background levels below one part per billion, being generally lowest in the southern autumn. *(NASA/JPL-Caltech)*

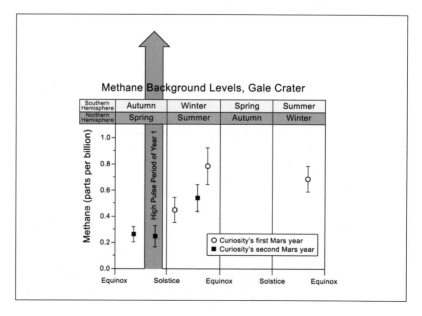

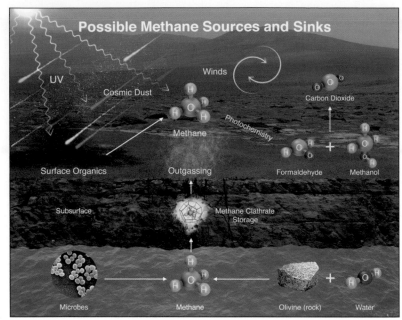

**Possible Methane Sources and Sinks**

nitrogen can be used by living organisms, this was further support for the idea that ancient Mars had been suitable for life.

On sol 746 Curiosity set about navigating a safe route through the field of dunes into the hills at the base of Aeolis Mons.

In early 2017 NASA proudly reported: "During the first year after Curiosity's 2012 landing in Gale Crater, the mission fulfilled its main goal by finding that the region once offered environmental conditions favourable for microbial life. Long-lived ancient freshwater lake environments had all of the key chemical elements needed for life as we know it, plus a chemical source of energy that is used by many microbes on Earth." But that lake had existed 3.3 to 3.8 billion years ago, and it would be

**ABOVE** Ways in which methane might be added to the atmosphere of Mars (sources) and removed from it (sinks). A molecule of methane has a single atom of carbon and four of hydrogen. It can be generated by microbes and by non-biological processes, for example in reactions between water and some rock minerals. Furthermore, the ultraviolet radiation from the Sun can induce reactions that generate methane from other organic chemicals resulting from either biological or non-biological processes, such as comet dust falling on Mars. Methane created underground might be stored in lattice-structured hydrates called clathrates and be released at a later time, so methane being released into the atmosphere today might have been formed in the distant past. Of course, the winds on Mars will rapidly distribute the methane from an individual source. Methane can be removed from the atmosphere by sunlight-induced reactions. These oxidise it by way of intermediary chemicals such as formaldehyde and methanol into carbon dioxide, which is the primary ingredient in the atmosphere. *(NASA/JPL-Caltech)*

**BELOW** A view by Curiosity toward the higher regions of Aeolis Mons. In the foreground, about 3km away, is a long ridge teeming with hematite. Beyond that is an undulating plain rich in clay minerals, and then a multitude of rounded buttes that are rich in sulphates. If Curiosity can sample these layers, the change of mineralogy will shed light on an evolving environment early in the planet's history. Farther back are striking, light-toned cliffs in rock that may have formed in drier times and now is heavily eroded by winds. Note that the colours are false in that they have been adjusted to make rocks look roughly as they would on Earth to help geologists interpret them. As a result of this 'white balancing' process the sky appears light blue. *(NASA/JPL-Caltech/MSSS)*

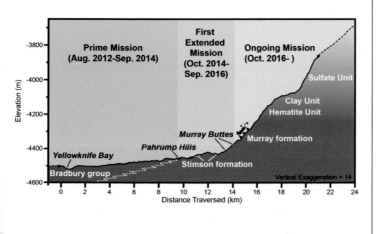

**TOP** In this view, the fact that the strata in the foreground dip towards the base of Aeolis Mons indicates a flow of water toward a basin which existed before the larger bulk of the mountain formed. *(NASA/JPL-Caltech/MSSS)*

**ABOVE** As Curiosity worked its way through the Murray Buttes it found itself looking up at rock structures. *(NASA/JPL-Caltech/MSSS)*

**LEFT** An illustration of the driving distance, elevation, geological units and time intervals of the Curiosity mission to December 2016, by which time it had gained 165m in elevation. Note that because Mars lacks a sea level, elevations are measured relative to an arbitrary datum. At Gale the elevations are negative. *(NASA/JPL-Caltech)*

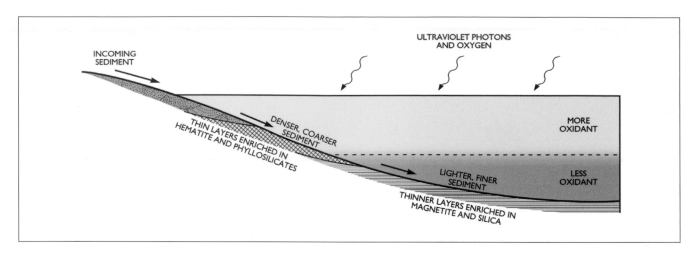

INCOMING
SEDIMENT

ULTRAVIOLET PHOTONS
AND OXYGEN

MORE
OXIDANT

LESS
OXIDANT

THIN LAYERS ENRICHED IN
HEMATITE AND PHYLLOSILICATES

DENSER, COARSER
SEDIMENT

LIGHTER, FINER
SEDIMENT

THINNER LAYERS ENRICHED IN
MAGNETITE AND SILICA

**ABOVE** Some of the processes and clues related to a long-ago lake on Mars that became stratified, with the shallow water richer in oxidants than the deeper water. The sedimentary rocks deposited within a lake in Gale crater more than 3 billion years ago differ from each other in a pattern which matches terrestrial lakes. As sediment-bearing water flows into a lake, bedding thickness and particle size progressively decrease as sediment is deposited in deeper and deeper water. At sites on lower Aeolis Mons, measurements of chemical and mineral composition by Curiosity revealed a clear correspondence between the physical characteristics of sedimentary rock from different parts of the lake and how strongly oxidised those sediments were. Rocks with textures indicating that the sediments were deposited near the edge of a lake have more strongly oxidised composition than rocks with textures indicating sedimentation in deep water. This chemical stratification occurs because the water closer to the surface is more exposed to the oxidising effects of oxygen in the atmosphere and solar ultraviolet light. On Earth, a stratified lake with a distinct boundary between oxidant-rich shallows and oxidant-poor depths provides a diversity of environments suited to different types of microbes. This provided a direct parallel for selecting locations to seek evidence of microbial life at Gale.

*(NASA/JPL-Caltech/Stony Brook Univ./Woods)*

**RIGHT** Curiosity's route of traverse. The blue star is the Bradbury Landing. Blue triangles mark waypoints investigated on the floor of Gale and (beginning with Pahrump Hills) on Aeolis Mons. The sol 1,750 label shows the location of the vehicle on 9 July 2017 (about the time of writing this text). At that time the mission planners were examining Vera Rubin Ridge from the downhill side. Spectrometry from Mars Reconnaissance Orbiter established the ridge to be rich in hematite. The planned route continues beyond the top of the ridge onto units where clay minerals and sulfate minerals have been detected from orbit. *(NASA/ JPL-Caltech/Univ. of Arizona)*

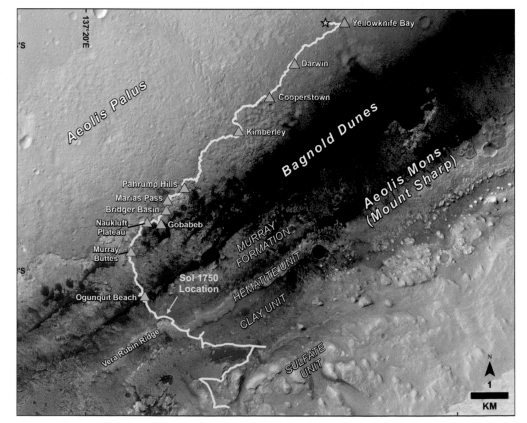

**RIGHT** The InSight lander that will use seismometers and heat flow probes to study the deep interior of Mars. *(NASA/JPL-Caltech)*

difficult to find unambiguous evidence in the rocks of life having existed then.

Several months later it was announced that the ancient lake in Gale crater had been 'stratified'. The shallows were rich in oxidants and the depths poor in oxidants, indicating that at any given time the lake would have offered different types of microbe-friendly environments.

At the time of writing this book, Curiosity is continuing its ascent of Aeolis Mons seeking evidence of how and when the early habitable environment became more stressful for any microbes that may have been present.

# Planetary core

In 2018 NASA is to launch the Interior Exploration using Seismic Investigations, Geodesy and Heat Transport (InSight) mission for the Discovery Program.

It will be a stationary lander similar to Mars Phoenix, but equipped to study the deep interior of the planet. A seismometer will determine the size, thickness, density, and overall structure of the core, mantle, and crust, and a heat transfer probe that digs into the ground to a depth of 5m will determine the rate at which heat is escaping from the interior. This data will help us to understand the early geological evolution of the planet.

These science goals do not impose any particular requirements in terms of surface feature. The site must be in the equatorial zone for efficient solar power all year round, be at a low elevation to maximise atmospheric braking at entry, be flat and relatively free of obstacles to reduce the likelihood of complications during landing, and soft enough to allow the heat flow probe to penetrate to the desired depth. As Elysium Planitia satisfies all these requirements a site in this area will be chosen.

**RIGHT** The interior of Mars is believed to consist of a crust, a mantle and a core. *(NASA/JPL-Caltech)*

# Future rovers

In 2020 NASA's Mars Exploration Program intends to use another skycrane to land a rover similar to the remarkable Curiosity, carrying instruments to further investigate an environment which appears likely to have been conducive to the development of life early in the planet's history. In particular, it will evaluate the potential for biomarkers preserved in materials that are accessible for sampling.

The plan is that the new rover will cache the most interesting of its samples for possible return to Earth by a subsequent mission.

In October 2016 the Schiaparelli entry, descent and landing demonstrator for the joint European-Russian ExoMars programme malfunctioned in the final stage of its descent. The real-time telemetry revealed the problem, which was fixable, so the mission was declared a partial success. If it had landed on Meridiani Planum the small science payload would have performed meteorological monitoring until its battery expired.

The next ExoMars mission is to be launched in 2020, and a Russian lander will deploy a European Space Agency rover. A variety of landing sites are being analysed and the final selection is expected approximately a year before launch.

The solar-powered rover is to spend at least six months seeking evidence of life on the planet in the past or indeed in current times.

It will drill to a depth of 2m at several locations to extract a number of core samples, each 3cm in length and 1cm in diameter. This plan is based upon the presumption that this depth will be sufficiently removed from the harsh surface conditions for micro-organisms that originated when Mars was a warm and wet world to survive to the present day.

As the rover explores, an external spectrometer will help to identify water-related minerals on the surface and a neutron detector will seek indications of subsurface hydration and the presence of buried water ice.

A ground-penetrating radar will map the shallow stratigraphy of the site and assist in selecting drill sites. A camera in the drill system

**BELOW** The rover that NASA intends to send to Mars in 2020 will use a drill to obtain core samples. Where the evidence shows conditions were habitable in the ancient past it will seek indications of past microbial life. *(NASA/JPL-Caltech)*

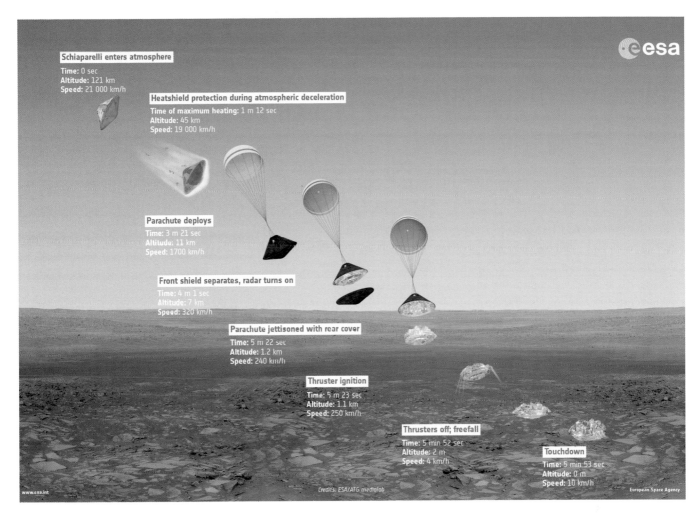

Schiaparelli enters atmosphere
Time: 0 sec
Altitude: 121 km
Speed: 21 000 km/h

Heatshield protection during atmospheric deceleration
Time of maximum heating: 1 m 12 sec
Altitude: 45 km
Speed: 19 000 km/h

Parachute deploys
Time: 3 m 21 sec
Altitude: 11 km
Speed: 1700 km/h

Front shield separates, radar turns on
Time: 4 m 1 sec
Altitude: 7 km
Speed: 320 km/h

Parachute jettisoned with rear cover
Time: 5 m 22 sec
Altitude: 1.2 km
Speed: 240 km/h

Thruster ignition
Time: 5 m 23 sec
Altitude: 1.1 km
Speed: 250 km/h

Thrusters off; freefall
Time: 5 min 52 sec
Altitude: 2 m
Speed: 4 km/h

Touchdown
Time: 5 min 53 sec
Altitude: 0 m
Speed: 10 km/h

www.esa.int

Credits: ESA/ATG medialab

European Space Agency

esa

will inspect the hole to observe the variation of water-related minerals with depth.

The core samples will be transferred inside the vehicle for distribution and analysis.

One instrument will seek organic molecules in the sample. It will operate in the same manner as the gas chromatograph/mass spectrometer of the Viking landers, but with much greater sensitivity. Another spectrometer that can identify mineral phases produced by water-related processes will seek evidence of life from its mineral products and other 'markers' indicative of microbial activity.

Some material will be crushed to a powder for examination by an infrared imaging spectrometer. Studying mineral grain assemblages will ascertain their origin, structure, and composition; essential evidence for interpreting past and present geological processes and environments on the planet.

The ExoMars rover will have an excellent chance of detecting organisms living in the subsoil – providing that they actually exist and

that our reasoning in selecting the landing and drilling sites is valid.

An unambiguous positive detection ought to resolve the question once and for all. Then there will be an almighty squabble to establish which previous data was, in retrospect, clear evidence of that fact. How the old data is prioritised will influence academic reputations and perhaps even the allocation of future Nobel Prizes!

Establishing that life is not unique to Earth would, as the late Philip Morrison of the Massachusetts Institute of Technology pointed out, "transform the origin of life from a miracle to a statistic".

If Mars proves never to have harboured life, the question becomes: why did the chemical evolution that led to terrestrial life not do so on Mars?

Irrespective of the verdict, the search for life on Mars addresses one of the most fundamental questions facing mankind.

**ABOVE The planned entry, descent and landing of the ExoMars Schiaparelli probe. It failed shortly after releasing its parachute and activating its braking thrusters.**
*(ESA/AGT Medialab)*

*Chapter Seven*

# Mars in fiction

Mars has featured prominently in science fiction due to its relative accessibility, the prospect of indigenous life, and how we might establish a colony there. We might even ultimately terraform the planet. Mars represents the next frontier to challenge the expansion of human presence.

**OPPOSITE** Artwork by Henrique Alvim Corrêa for the 1906 Belgian translation of the Herbert George Wells novel *The War Of The Worlds*.

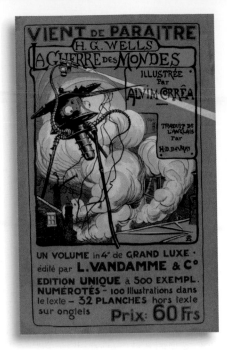

# The War of the Worlds

After returning from the Far East to Boston in Massachusetts in 1893, Percival Lowell, who had an interest in astronomy, was given a copy of *The Planet Mars* as a gift. In this heavy tome, published in France the previous year, astronomer and prolific author Camille Flammarion discussed the history of observations of the planet right through to the recent reports by Giovanni Virginio Schiaparelli in Italy of linear features named 'canali'.

After devouring the book, Lowell built an observatory near Flagstaff in what was then the Arizona Territory to make a personal study of Mars during its 1894 opposition. In 1895 he wrote up his observations in a book entitled simply *Mars*, along with his thoughts about conditions on the planet. He argued that the canali were created by an ancient race of beings in an effort to irrigate their arid planet.

Lowell's vision of Mars as a dying world prompted Herbert George Wells in England to write *The War of the Worlds*, a story in which Earth is invaded by a warlike race of creatures from Mars.

It was serialised in *Pearson's Magazine*, a monthly that began publishing in 1896 and featured what it called 'speculative' forms of literature. It carried Wells' story from April to December 1897. The next year it was published as a novel by William Heinemann. It was an instant best-seller, and has never been out of print.

This paragraph opened the story:

*"No one would have believed in the last years of the nineteenth century that this world was being watched keenly and closely by intelligences greater than man's and yet as mortal as his own; that as men busied themselves about their various concerns, they were scrutinised and studied, perhaps almost as narrowly as a man with a microscope might scrutinise the transient creatures that swarm and multiply in a drop of water. With infinite complacency men went to and fro over this globe about their little affairs, serene in their assurance of their empire over matter. It is possible that the infusoria under the microscope do the same. No one gave a thought to the older worlds of space as sources of human danger, or thought of them only to dismiss the idea of life upon them as impossible or improbable. It is curious to recall some of the mental habits of those departed days. At most, terrestrial men fancied there might be other men upon Mars, perhaps inferior to themselves and ready to welcome a missionary enterprise. Yet across the gulf of space, minds that are to our minds as ours are to the beasts that perish, intellects vast and cool and unsympathetic, regarded this Earth with envious eyes, and slowly and surely drew their plans against us."*

To enhance the plausibility of his storytelling Wells made explicit mention of the latest scientific thinking, including the names of people such as Schiaparelli.

Wells told of the Martians landing where he lived, west of London, and then related their assault on the city. For the invasion, they had encased themselves inside indestructible machines possessing tripod legs. They had beam weapons to slaughter people mercilessly. Just as they were on the verge of winning, they were struck down by a bacterial infection.

Interestingly, Wells made no attempt to characterise the aliens; they were simply monsters.

Late Victorian society was both shocked and thrilled by the brutality of the invaders.

In fact Wells was telling an allegory. At that time Britain possessed a global Empire "upon which the Sun never set". In acquiring territories, the British Army had simply invaded, pushed aside the natives, and imposed their own laws and institutions. Wells was presenting a version of this story from the point of view of the oppressed; in this case the whole of humanity.

Lowell further expounded upon his vision in 1906 with *Mars and its Canals* in which he painted a more complete picture of the valiant effort of the inhabitants to prevent being overwhelmed by a process of desertification. He wrote, "That Mars is inhabited by beings of some sort we may consider as certain as it is uncertain what these beings may be."

Then in 1908, in *Mars as the Abode of Life*, he discussed how the different conditions existing on Earth and Mars must have driven the development of life and influenced their intelligent races.

## Barsoom and John Carter

Edgar Rice Burroughs in America adopted Lowell's vision of conditions on Mars and dreamed up his own inhabitants.

Whereas Lowell had imagined the Martians as benign as a result of working together to survive on a dying planet, and Wells had made them warlike to tell his story of brutal invasion, Burroughs produced a romantic adventure. And whereas Wells sought to impart scientific plausibility in his writing, Burroughs conceived a fantasy world and populated it with interesting characters.

Transported to Mars by a mysterious process, John Carter finds the planet inhabited by two intelligent races, one nomadic and the other city dwellers, who are forever at war. Burroughs

referred to the planet as Barsoom since that was the native name for it. During his adventures Carter falls in love with a beautiful princess.

The story was written in 1911, then serialised by a monthly pulp magazine called *The All-Story* from February to July 1912 under the pen-name of Norman Bean. In 1917 it was released as a book called *A Princess of Mars* under the author's own name. In later years, Burroughs added a number of sequels that comprise the *John Carter on Mars* series.

**ABOVE** Edgar Rice Burroughs.
*(Britannica.com)*

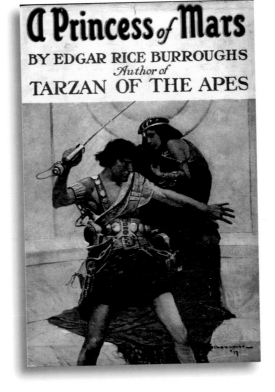

**LEFT** The cover of the 1917 release of *A Princess of Mars* by Burroughs featured artwork by Frank E. Schoonover.

His tale reinforced the popular impression that Mars was habitable, if not actually inhabited in the manner depicted.

## Radio scare

In 1938 the Columbia Broadcasting System in New York transmitted a series of radio dramas produced by the Mercury Theatre on the Air.

On 30 October, to mark Halloween, it aired *The War of the Worlds* adapted by Howard Koch and John Houseman. It was produced, directed and narrated by Orson Welles. Since they were aiming at an American audience, they moved the story to New Jersey.

The show ran for an hour without pauses for sponsorship. It started off as a normal broadcast featuring music. After a few minutes Welles interrupted with a news flash that astronomers were reporting bright flashes on Mars. Shortly after that he followed up with a report of a strange object falling to Earth close to Grover's Mill, which really existed. The broadcast then became a series of ever more dramatic 'live' reports of monsters spreading death and destruction.

At that time people had their radios on in the background in much the same way as a later generation would with its televisions.

**ABOVE Orson Welles at the Mercury Theatre on the Air in 1938.**
*(CBS)*

**RIGHT Orson Welles orchestrates the radio play based on *The War of The Worlds* that was broadcast by the Mercury Theatre on the Air on the eve of Halloween in 1938.**
*(CBS)*

**FAKE RADIO 'WAR' STIRS TERROR THROUGH U.S.**

"War" Victim

"I Didn't Know"

**Radio Listeners in Panic, Taking War Drama as Fact**

*Many Flee Homes to Escape 'Gas Raid From Mars'—Phone Calls Swamp Police at Broadcast of Wells Fantasy*

A wave of mass hysteria seized thousands of radio listeners throughout the nation between 8:15 and 9:30 o'clock last night when a broadcast of a dramatization of H. G. Wells's fantasy, "The War of the Worlds," led thousands to believe that an interplanetary conflict had started with invading Martians spreading wide death and destruction in New Jersey and New York.

The broadcast, which disrupted households, interrupted religious services, created traffic jams and clogged communications systems, was made by Orson Welles, who as the radio character, "The Shadow," used to give "the creeps" to countless child listeners. This time at least a score of adults required medical treatment for shock and hysteria.

In Newark, in a single block at Heddon Terrace and Hawthorne Avenue, more than twenty families rushed out of their houses with wet handkerchiefs and towels over their faces to flee from what they believed was to be a gas raid. Some began moving household furniture.

Throughout New York families left their homes, some to flee to near-by parks. Thousands of persons called the police, newspapers

and radio stations here and in other cities of the United States and Canada seeking advice on protective measures against the raids.

The program was produced by Mr. Welles and the Mercury Theatre on the Air over station WABC and the Columbia Broadcasting System's coast-to-coast network, from 8 to 9 o'clock.

The radio play, as presented, was to simulate a regular radio program with a "break-in" for the material of the play. The radio listeners, apparently, missed or did not listen to the introduction, which was: "The Columbia Broadcasting System and its affiliated stations present Orson Welles and the Mercury Theatre on the Air in 'The War of the Worlds' by H. G. Wells."

They also failed to associate the program with the newspaper listing of the program, announced as "Today: 8:00-9:00—Play: H. G. Wells's 'War of the Worlds'—WABC." They ignored three additional announcements made during the broadcast emphasizing its fictional nature

Mr. Welles opened the program with a description of the series of

**Continued on Page Four**

**FAR LEFT** The next day, the *New York Daily News* gave front-page coverage to the panic created by the radio broadcast.

**LEFT** Similar coverage in *The New York Times*.

Network radio was a trusted source of news. And in October 1938 the news was downbeat with the prospect of Nazi Germany invading its neighbours. Weeks earlier the immediate crisis had been defused by the signing of an agreement in Munich that seemed to appease Adolf Hitler's demands.

Residents of the small rural settlement grabbed their guns and congregated to defend their community from the invaders. Of course, portable transistorised radios were still a thing of the future, so as soon as people left their homes they lost touch with the play. But this didn't matter because, as happens all too easily in a panicky situation, those involved spread their own rumours of terrible events underway. As a result, after the show concluded by wishing its listeners a happy Halloween, the panic continued through the night.

In the aftermath the newspapers had a field day!

What this incident showed was that the public, in America at least, were all too willing to believe in the existence of Martians.

## The Martian Chronicles

The John Carter saga by Edgar Rice Burroughs influenced fellow American Ray Bradbury to write a number of short stories

in the late 1940s that were published in 1950 as *The Martian Chronicles*. The book was a roaring success.

Bradbury retained the Lowellian vision of Mars as an arid world with canals providing irrigation, and populated it with a race of benign, semi-physical, semi-spiritual beings who communicated telepathically.

**LEFT** A monument at the real hamlet of Grover's Mill, New Jersey, where the radio broadcast started off its version of the invasion from Mars. *(Courtesy of ZeWrestler, Wikipedia)*

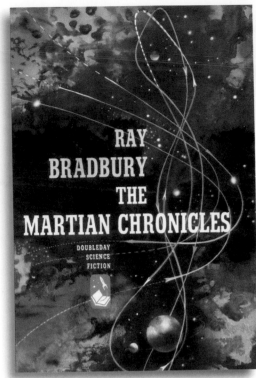

In essence, Bradbury inverted Wells by having the humans cross space to Mars and play the part of the unwelcome visitors. Furthermore, unlike Wells and Burroughs who set their stories in contemporary times, Bradbury imagined his in the future.

Bradbury begins with the first attempts by human expeditions to reach Mars at the end of the 20th century, with the Martians preventing them from returning home. But then the Martians fall victim to a human disease which almost wipes them out. Later, humans settle what appears to be an empty planet and start to make it into a second Earth, occasionally encountering some surviving Martians. However, with war looming on Earth, most of the settlers return home. When the war ends in nuclear holocaust, the last remaining humans are the ones living on Mars. They become the new Martians.

*The Martian Chronicles* had a major influence on later authors of what was now called 'science fiction'.

## Red Planet

In the late 1940s Robert Anson Heinlein published a number of novels aimed at juveniles in the sense that their main characters were teenagers. Nevertheless, the science fiction was first class. *Red Planet*, released in 1949, is considered a classic.

The setting is Lowellian, with deserts and canals in which the water freezes during the night. There are ancient cities and Martians who tolerate the presence of human settlers, but tend to remain aloof. Colonists migrate annually to and fro between hemispheres to avoid the harsh winters. Attending Lowell Academy, two youngsters find that the company which runs the planet has been postponing the migration of their home settlement in an attempt to force it to remain in place as a cost-saving venture, even if that means some of the settlers will perish. The boys escape and save the day with the help of the Martians, who are revealed to have an underground transportation system.

The twist in the story is that a small intelligent creature that one of the boys has adopted as a pet will soon metamorphose into an adult Martian. On learning of the intention of the manager of the company to send this creature to a zoo on Earth, the Martians destroy him by a mysterious process, then order the humans to vacate the planet. But the friendship between the boys and the young Martian sways the decision in favour of allowing the settlers to remain. It is also revealed that when adult

Martians die they turn into 'Old Ones' who inhabit another plane of existence, a theme that the author would pursue further in later works.

Although a visionary tale, *Red Planet* was considerably less ambitious than the broad canvas on which Bradbury painted.

## The Sands of Mars

*The Sands of Mars* was Arthur C. Clarke's first science fiction novel. Published in 1951. it is a strictly nuts and bolts account of humans establishing a settlement on Mars.

The planet is a desert and the thin atmosphere is sufficient to enable people to work outside with breathing apparatus. Although there is vegetation there was never a Martian civilisation to create a canal network.

A small colony has been established in a dome at Port Lowell, but the Earth authorities do not consider it worth the exorbitant cost and are thinking of calling everyone home.

The story begins with an elderly journalist travelling from Earth to Mars on a resupply vessel. During the voyage he befriends a young officer. While travelling around Mars the two men happen across a secret facility that is adapting a plant to release oxygen. It is intended to enrich the oxygen content of

RIGHT The 1951
edition of Clarke's
*The Sands of Mars*
featured artwork by
Chesley Bonestell.

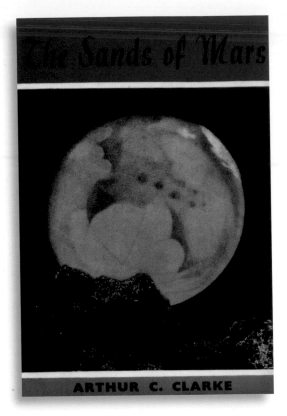

the atmosphere. Another part of the project involves turning the moon Phobos into a second Sun, as a way to warm the planet. When the latter is achieved, it radically

transforms the long-term prospects for colonising the planet. With a *fait accompli*, Earth will have to expand operations. The journalist decides to become a settler.

With *The Sands of Mars*, storytelling about Mars had moved from fantasy with Burroughs, through visionary with Heinlein, to 'hard' science fiction.

It is significant that Clarke, accepting what at that time were believed to be surface conditions on Mars, decided to have his colonists set about making the planet more hospitable.

## Space Age fiction

With the arrival of the Space Age, Mars was shown to be very different to what the best-informed scientists had believed.

It was an arid and very cold world possessing a very rarefied atmosphere. There were no seabeds lined with vegetation, no intelligent Martians, and no network of canals. Instead there was an astonishing landscape of craters, dry river valleys, vast outflow channels, and enormous volcanoes. Yet the surface was ancient, dating back billions of years.

Although Mars had evidently been warm and wet early in its history, it had indeed undergone a process of desertification and died, but not in the way that Percival Lowell had imagined it.

The revelation of Mars as it is, rendered obsolete the many tales depicting Martian civilisations. It even undermined what Arthur C. Clarke had reckoned to be a far more realistic scenario.

But that didn't prevent science fiction writers from telling tales about Mars. They simply accepted the situation as it was and then extrapolated from there.

One notable example was *Voyage* by Stephen Baxter, published in 1996. This portrays how the first mission to Mars might have been carried out by the expedient of modifying the technology created for the Apollo lunar missions.

Much more ambitious was the award-winning saga in 1,700 pages by Kim Stanley Robinson issued as *Red Mars* in 1992, *Green Mars* in

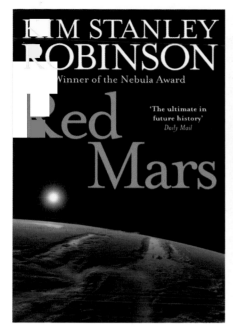

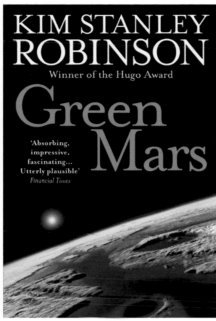

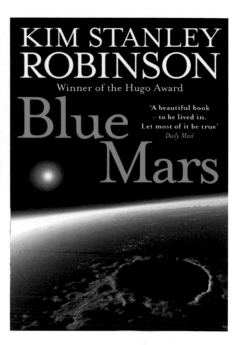

1993 and *Blue Mars* in 1996. The depiction of landscapes is enhanced by reference to maps compiled from orbital imagery.

Like Clarke but on a grander scale, Robinson's colonists embark on a 200-year project to transform Mars into a world better suited to humans. In reality of course, such a grandiose project lies far in the future.

When and how will humans really venture to Mars?

**ABOVE The *Mars* trilogy by Kim Stanley Robinson.**

## COMMUNICATING WITH MARTIANS

For much of the time that astronomers were studying Mars through telescopes they thought it only natural that the planet must be inhabited.

For example, in the late 18th century in writing up his observations of the planet, William Herschel casually mentioned that its inhabitants were probably enjoying conditions similar to himself.

A century later, when Schiaparelli drew linear streaks on Mars and Lowell claimed that they were artificial, there was nothing particularly shocking about this.

In 1900 the Académie des Sciences in Paris announced it would award a prize of 100,000 francs to the first person to open communications with aliens. The money was pledged by Clara Goquet Guzman in memory of her son, who had been fascinated by astrology. Although she specifically excluded Mars for being too easy a target, soon people, most notably Nikolai Tesla, claimed they had received radio signals from the planet.

In 1909 William Henry Pickering, an associate of Lowell, recommended the construction of thousands of mirrors in order to signal to the inhabitants of Mars.

Some years earlier, Charles Cros, a colleague of Camille Flammarion, had petitioned the French government to construct a gigantic mirror to communicate with Martians by burning giant lines on the desert terrain of their planet. Such an action could hardly have been considered a friendly act by the inhabitants!

A less aggressive idea for the particularly favourable opposition of 1956 was to place giant symbols in the Sahara desert for the Martians to see when the two planets were at their closest. The flaw was that when Mars was high in the sky at midnight, the Sahara was in darkness! In any case, to a Martian observer at that time, Earth would be close to the Sun in the sky and hence impossible to inspect with the requisite attention to detail.

*Chapter Eight*

# When will humans visit Mars?

In the early 1950s studies were made of flights to Mars. The early concepts were naïve but they led to the development of better ones. There was the possibility of adapting Apollo technology for a mission in the 1980s, but this was not pursued. We should soon be able to mount such a journey with new vehicles and we face decisions about the objectives.

**OPPOSITE A depiction by Pat Rawlings of a human expedition to explore the chasms of Noctis Labyrinthus.** *(NASA)*

## Space travel

**P**rior to the Space Age, those who told stories of humans on Mars tended not to dwell on how they got there.

Edgar Rice Burroughs, for example, simply had his hero wish himself to be on Mars and suddenly he was on the planet. Ray Bradbury and Robert Heinlein acknowledged rocket travel, but in *The Sands of Mars* Arthur C. Clarke explicitly featured the voyage.

Meanwhile, technical developments were making rocket propulsion a reality, most notably in the form of the V-2 missile that was introduced by Nazi Germany during World War II. A key player in that effort was the young engineer Wernher von Braun. Immediately after the war he and many of his colleagues were taken to America to continue their work for the US Army.

It was a rocket developed by von Braun by improving on the V-2 design that placed America's first satellite into orbit in January 1958. It carried an instrument supplied by James van Allen that discovered a belt of charged particle radiation encircling the planet.

When in 1961 President John Fitzgerald Kennedy challenged his nation to "land a man on the Moon" within the decade, von Braun led the development of the mighty Saturn V rocket that sent the Apollo spacecraft Moonward.

Immediately upon the achievement of Kennedy's challenge in 1969, there were calls for a similar 'all out' effort to reach Mars by 1980.

In fact, von Braun had already described how such an expedition could be carried out.

## An early vision

**I**n 1953 the University of Illinois Press published a book by von Braun entitled *The Mars Project*. It was an English version of a piece he penned the previous year for the German magazine *Weltraumfahrt* (*Spaceflight*) as *Das Marsprojekt*.

Von Braun explained that most of the cost of a space expedition, at least in terms of energy expenditure, would be lifting things off Earth into orbit. Hence it made sense to assemble vessels intended for interplanetary travel in space. In that way, they need not be constructed to withstand the rigours of gravity or be aerodynamically shaped. He envisaged three-stage winged rockets carrying all the hardware and propellant into orbit, where the interplanetary ships would be assembled. In addition to the ships optimised for use in space, there would also be 'boats' for landing on Mars.

**RIGHT** The cover of the 1952 German edition of Wernher von Braun's *Das Marsprojekt*.

**FAR RIGHT** Wernher von Braun in 1952 with a model of the three-stage rocket that he envisaged would ferry large payloads into orbit, where they would be assembled for a mission to Mars. *(NASA)*

The expedition would involve a fleet of ten ships, each carrying a crew of seven (all of them male, of course). As each craft would be 3,000 tonnes fully laden, almost 1,000 launches would be needed to initiate an expedition.

Once in orbit around Mars, the first landing boat would glide down to one of the polar caps and land on the ice using skids. After deploying vehicles, its crew would trek several thousands of kilometres to the equatorial zone to construct a runway on which the other two boats with wheels would set down to deliver the majority of the crew (some would remain behind to look after the ships in orbit). They would erect inflatable habitats and live on the planet for 400 days or so. In addition to a programme

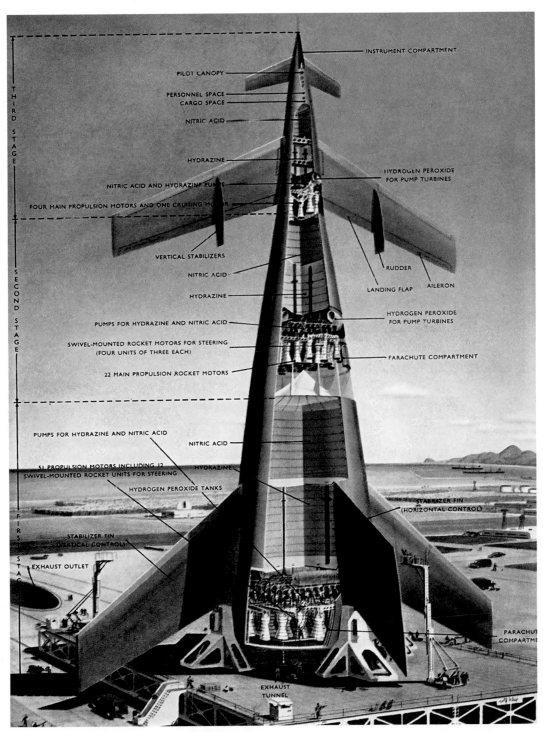

INSTRUMENT COMPARTMENT

PILOT CANOPY

PERSONNEL SPACE
CARGO SPACE

NITRIC ACID

HYDRAZINE

HYDROGEN PEROXIDE
FOR PUMP TURBINES

NITRIC ACID AND HYDRAZINE PUMPS

FOUR MAIN PROPULSION MOTORS AND ONE CRUISING MOTOR

THIRD STAGE

VERTICAL STABILIZERS

NITRIC ACID

RUDDER

HYDRAZINE

LANDING FLAP

AILERON

SECOND STAGE

PUMPS FOR HYDRAZINE AND NITRIC ACID

HYDROGEN PEROXIDE
FOR PUMP TURBINES

SWIVEL-MOUNTED ROCKET MOTORS FOR STEERING
(FOUR UNITS OF THREE EACH)

PARACHUTE COMPARTMENT

22 MAIN PROPULSION ROCKET MOTORS

PUMPS FOR HYDRAZINE AND NITRIC ACID

NITRIC ACID

51 PROPULSION MOTORS INCLUDING 12
SWIVEL-MOUNTED ROCKET UNITS FOR STEERING

HYDRAZINE

HYDROGEN PEROXIDE TANKS

STABILIZER FIN
(HORIZONTAL CONTROL)

FIRST STAGE

STABILIZER FIN
(VERTICAL CONTROL)

EXHAUST OUTLET

PARACHUTE
COMPARTMENT

EXHAUST
TUNNEL

**LEFT Cutaway artwork by Rolf Klep of von Braun's ferry rocket.** (Collier's)

of scientific research, they would raise the two boats to a vertical orientation and remove their wings to eliminate 'dead weight' when they lifted off to return to their motherships.

Von Braun's book had two obvious merits. It explained how an expedition to Mars could be undertaken, and it provided a mathematical explanation to enable people to appreciate the complexity of the operation.

Of course some aspects of the proposal would prove to be implausible as a result of Mars having an atmosphere only about one-tenth the pressure that von Braun had presumed (on the best scientific knowledge at the time). A winged landing boat would therefore be unfeasible. And the surface of Mars isn't a vast, flat plain (as astronomers believed), it is a very rough landscape and driving from pole to equator to construct a runway would not have been realistic for any reasonable mission timescale. Also, because its existence had not yet been discovered, von Braun gave no thought to the risks of space radiation.

Nevertheless, *The Mars Project* served its purpose by defining what would nowadays be called a 'reference mission' as the basis for further investigation.

It was introduced to the American public by the popular New York monthly magazine *Collier's* in 1954, in an article written by von Braun and illustrated by Chesley Bonestell.

In *The Exploration of Mars*, published in 1956, von Braun and space writer Willy Ley greatly simplified the concept.

Firstly they cut the number of interplanetary ships to two and the total crew complement to 12 men. Then they dispensed with the polar landing, the trek to the equator, and the construction of a runway. Instead, after the ships achieved orbit around Mars a telescopic inspection would reveal a flat, sandy plain in the equatorial zone that would serve as a natural runway. An unmanned craft would verify the suitability of the site, then parachute down with a radio beacon for the manned boats, whose wheels would have soft tyres for unprepared terrain. The tasks for the nine-man surface team would include preparing the boats for eventual lift-off, as in the previous plan.

But despite the historic first landing of men on the Moon in July 1969, the technology available fell far short of what von Braun had imagined for flying to Mars. With no prospect of developing such vehicles any time soon, it was

**RIGHT** Chesley Bonestell with one of his iconic paintings, a globe of Mars and some model rockets. *(Collier's)*

**FAR RIGHT** The cover of the 30 April 1954 issue of *Collier's* magazine featuring von Braun's vision of a mission to Mars.

clear that if Mars were to be reached by 1980, it would be necessary to use the Saturn V and substantially improve the capabilities of the Apollo spacecraft, at least by adding a habitat to sustain the crew in space for several years.

## A post-Apollo proposal

On 4 August 1969, with the Apollo 11 crew still in quarantine after their return to Earth with lunar samples, von Braun gave a presentation to the White House's Space Task Group.

Chaired by Vice President Spiro T. Agnew, the STG was to recommend the major objectives that NASA should pursue in the coming decade. One possibility was to spend that time developing the vehicles necessary for a mission to Mars. Von Braun indicated how such a project might be carried out based on what had recently been discovered about the planet, most notably that its atmosphere was extremely rarefied.

**ABOVE Depictions of the simplified 1956 version of von Braun's mission to Mars.** *(Collier's)*

**LEFT Wernher von Braun with the mighty Saturn V that would launch the Apollo 11 mission to attempt the first human landing on the Moon.** *(NASA/KSC)*

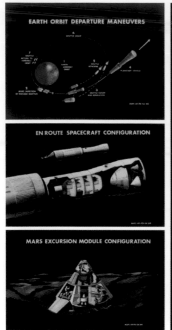

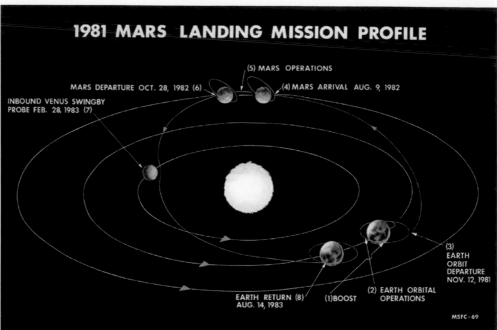

## 1981 MARS LANDING MISSION PROFILE

EARTH ORBIT DEPARTURE MANEUVERS

EN ROUTE SPACECRAFT CONFIGURATION

MARS EXCURSION MODULE CONFIGURATION

(5) MARS OPERATIONS

(4) MARS ARRIVAL AUG. 9, 1982

MARS DEPARTURE OCT. 28, 1982 (6)

INBOUND VENUS SWINGBY
PROBE FEB. 28, 1983 (7)

(3)
EARTH
ORBIT
DEPARTURE
NOV. 12, 1981

EARTH RETURN (8)
AUG. 14, 1983

(1) BOOST

(2) EARTH ORBITAL
OPERATIONS

MSFC-69

**ABOVE Slides from von Braun's 1969 presentation to NASA outlining how a mission to Mars could be mounted in the early 1980s.** *(NASA)*

Gone were the winged landing boats and the runways. Now the landings would be made by small vehicles (modules) that would descend vertically with rocket engines facilitating a soft landing.

Two interplanetary ships based upon Apollo technology would leave Earth, each carrying a crew of six men. They would fly in formation so that if one were to suffer difficulties its crew would be able to transfer to the other ship and continue the mission. Von Braun specified a launch in November 1981 because it was the first window of that decade. On achieving orbit around the planet in August 1982 six men would land, three in each lander, and spend two months on science tasks. Then they would join their colleagues in orbit. The ships would depart in October 1982, pursue a return trajectory that would make a 'gravity assist' flyby of Venus and arrive home in August 1983.

Von Braun noted that further missions could be flown at successive launch windows to build towards a 15-man base by the end of the decade.

But the possibility of adapting Apollo technology to reach Mars fell by the wayside when it was decided to pursue the development of a reusable space vehicle. It was expected that among its many tasks, this Space Shuttle would allow a renewal of human lunar operations and open up targets such as

Mars. Unfortunately, it limited human operations to low Earth orbit well into the 21st century.

Nevertheless, engineers continued to study a variety of alternative ways of reaching Mars in a more affordable manner.

## What, how, and when

On the retirement of the Space Shuttle in 2011, the prospects for a mission to Mars became more positive, although it was acknowledged that nothing would happen any time soon.

So what are the options?

An analogy can be drawn with Apollo.

In early 1968 the intention was to fly a command module on its own in low Earth orbit as Apollo 7 in the autumn, then fly it jointly with the lunar module in that environment by the end of the year as Apollo 8. In 1969, Apollo 9 would fly the two vehicles in an orbit with a high apogee, possibly even to lunar distance without the Moon being present, in order to test operating in deep space. Then Apollo 10 would go into lunar orbit to rehearse procedures there, except for the lunar module initiating a powered descent to the lunar surface. That would pave the way for Apollo 11 to attempt to land.

However, in the autumn of 1968 the development of the lunar module was running behind schedule and rather than ground Apollo 8 awaiting delivery of its lunar module NASA decided to send the command module into lunar orbit. This was an audacious and risky move but it offered early experience of operating in deep space and lunar orbit. When a lunar module became available in the new year, Apollo 9 tested it in low Earth orbit and the later missions flew as planned.

It may sound reasonable that having travelled all that way to Mars, the first human mission should attempt a landing. But that would have corresponded to NASA requiring the first Apollo crew to reach lunar orbit also to attempt to land. Only in science fiction would such a venture have been logical. In reality, it was recognised that operating around the Moon would be very different to low Earth orbit and that precursor flights would provide the opportunity to test technology and procedures. The need to learn lessons will be even more true when we first venture into deep space as far as Mars.

A mission that was required only to orbit Mars would be much simpler than one whose objective was to land on the planet. Furthermore, by not requiring the lander, an orbital mission could be flown much earlier to test all of the other components, particularly the deep space habitat that would be required to sustain the crew for several years.

An orbital mission could carry out useful science, such as visiting the small Martian moons and sending rovers down to the planet to collect samples which would then be lifted into orbit, thereby making the first human presence in Mars orbit a sample-return mission.

To enable humans to reach the surface of Mars it will be necessary to create a lander that is capable of carrying loads far heavier than even the largest robotic rovers. Prior to taking the crew down from orbit, such vehicles will have to deliver a surface habitat, surface vehicles, and a miscellany of supplies. A mission to the surface will therefore be a major undertaking. And of course the crew will have to be able to lift off and rendezvous with the mothership in orbit.

If the strategy calls for the first human expedition that achieves orbit around Mars to land on the planet, then that will not be able to be attempted until all the specialised hardware has been developed and tested, which by any reasonable projection is many years away. But a precursor in the style of Apollo 8 that was required only to operate in orbit around Mars could be undertaken much sooner and could give the programme a phased implementation that would allow valuable lessons to be learned in advance of attempting to land on the planet.

The view of any given individual on this topic will depend on their personal background and motivation. Some scientists are appalled at the prospect of an expensive human mission to Mars, arguing instead that we need more robotic landers and rovers. Other scientists fancy themselves as explorers who will be selected for the first crew to land on the planet. Indeed, some claim they would accept a one-way trip! It is interesting that those urging caution tend to be the engineers familiar with the challenges of developing procedures and fall-back options for flying missions in space.

In the decades since von Braun outlined how a landing on Mars might be achieved using Apollo-era technology a vast number of studies have explored other options.

To overcome the challenge of a massive lander capable of carrying a crew and all of the facilities they would require on the surface, many proposals called for the cargo to be flown to Mars by a different spacecraft (possibly of a different type) from that carrying the crew. This would allow a cargo ship to be launched at an earlier launch window.

An innovative concept recommended the strategy of 'living off the land' by exploiting resources on Mars to manufacture some of the consumables for the mission. In particular, it would be possible to send a lander to manufacture the propellant that the lander for a human mission would require to lift off. Doing so would further reduce the mass of the crewed lander. Obviously the crew would not commit to a landing until they knew the propellant for their return to orbit was available.

Another way to reduce mass was not to carry propellant from Earth to use during braking manoeuvres. Instead, aerodynamic interaction with the planet's atmosphere would be used when decelerating into orbit upon arrival and again when landing.

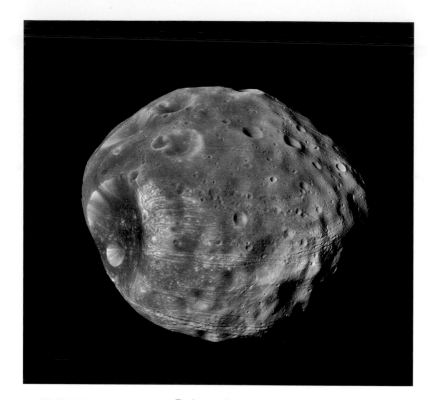

**ABOVE** Phobos, the larger of the two moons of Mars, as imaged in 2008 by Mars Reconnaissance Orbiter. *(NASA/JPL-Caltech/Univ. of Arizona)*

Although this was a rather minimalist strategy in terms of development time and cost, further missions could follow at successive launch windows to gradually build up a base of operations on Mars.

Another concept for a sustained effort envisaged launching a 'cycler' vehicle into a solar orbit. As its name implies, by making a succession of flybys of both Earth and Mars this would enable crews to travel to and fro cheaply. The cycler would depart the vicinity of Earth for a six-month journey to Mars every 20 to 30 months. People heading for Mars would leave Earth orbit on a small shuttle and rendezvous with the cycler. At Mars they would separate from the cycler in that vehicle, aerobrake into orbit around the planet, and land. The outpost on Mars would be a permanent base occupied by a succession of crews. This concept is being championed by Apollo 11 astronaut Buzz Aldrin.

Fewer studies have been made into how an orbital mission might be flown, basically because it was taken for granted that a crew heading for Mars would wish to land on the planet.

In scientific terms, an orbital mission offers the chance to investigate the Martian moons, which would probably be neglected by a crew charged with a planetary landing because the operational requirements for a mission that will operate in the vicinity of small moons are very different to one focusing on the planet. And switching the action to the moons will require a small spacecraft to enable astronauts to exit

Perhaps the most influential study to employ this kind of thinking was 'Mars Direct'. Proposed by Robert Zubrin in 1991, the rationale was to demonstrate that a human Mars mission could be affordable. Rather than assume the introduction of advanced rocketry, he opted for conventional chemical propulsion, factored in direct launch from the surface of Earth (hence the name), aerodynamic braking at Mars, production of propellant in situ, and direct return from the surface of Mars, along with sending out essential elements in advance.

**RIGHT** Mars Reconnaissance Orbiter imaged Deimos in 2009. *(NASA/JPL-Caltech/Univ. of Arizona)*

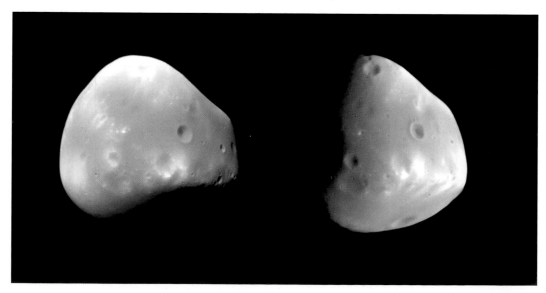

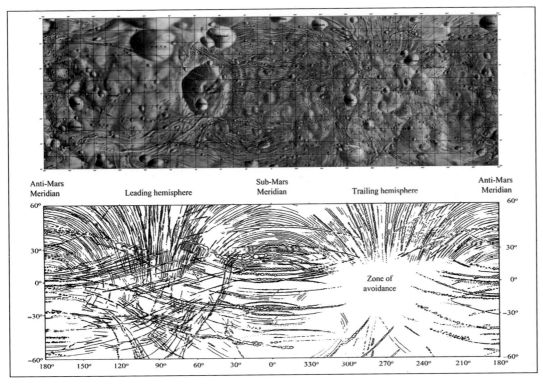

**LEFT** Grooving on Phobos. The map projection is courtesy of Philip Stooke and the plot of grooves, based on imagery from the High Resolution Stereo Camera of the ESA Mars Express spacecraft, was published by J.B. Murray and J.C. Iliffe as 'Martian Geomorphology' with the Geological Society of London, 2011.

their mothership and fly across to a moon to conduct science activities.

Propellant is a precious resource when it needs to be carried from Earth. It can all too easily be consumed by an elaborate programme of manoeuvres to visit moons. The Soviet Fobos 2 spacecraft spent a lot of time setting up a series of close flybys in preparation for releasing small landers. However, a sequence of brief encounters would not be appropriate for a human mission focusing on the moons. It would be preferable to manoeuvre close to a moon and study it for a prolonged period. Fortunately, this is possible.

A technique first described by the 18th century Italian mathematician and astronomer Joseph-Louis Lagrange can be used to reduce the expenditure of propellant in a so-called 'three body problem'.

In 1772, while studying how the gravitational fields of the Earth, Sun, and Moon interact, Lagrange discovered a number of special case solutions which offer stability. The analysis applies to any situation in space where the central body is much larger than one that is orbiting around it, with a third that is even smaller in the vicinity. It therefore applies to Mars, a moon of that planet and a spacecraft.

For such a system there are five locations in space, called the Lagrangian points. For a spacecraft in Mars orbit, the L1 point will be situated between the planet and the moon in question, the L2 point will be on the opposite side of the moon, and the L3 point will be behind the planet. So they are all in a line which passes through the centres of the two bodies and rotates around the planet with the moon. The L4 and L5 points occupy the same orbit as the moon and are 60° ahead of and in trail of it respectively.

Interestingly, the interacting gravitational fields of the planet and moon give the impression of a 'locus of attraction' being at a Lagrange point. As a result, a spacecraft that is manoeuvred to a Lagrange point will tend to stay in place. Of course perturbations will cause it to drift, but at least in the short term these can be countered by propulsive burns. This is called 'station keeping'. In the case of Phobos and Deimos, the moons are so small that the propellant expenditure to stay in position at a Lagrange point will be negligible.

To investigate a Martian moon, the mothership could be manoeuvred into either the L4 or L5 point and some members of the crew would fly to the moon using a subsidiary craft.

As both tiny moons are tidally locked, meaning that they maintain one face toward Mars, a craft at a Lagrangian point will remain

at a fixed point above the surface. Since the L1 and L2 points of Phobos are only a few kilometres away, this readily suggests two possibilities for obtaining samples. A collection device could be released, unreeling a tether. When it had obtained a surface sample it would be hauled up. Alternatively a tethered harpoon could act as an anchor to enable the spacecraft to drag itself down to the surface so that astronauts could venture outside to collect samples or emplace instruments.

Once the surface activities were complete, the small vehicle would return to the mothership, which would then nudge itself out of its own Lagrange point and pursue the next phase of the mission.

The order in which the moons would be visited would depend on a variety of mission constraints and the relative scientific priorities assigned to the individual moons.

This activity involving the moons would be distinct from operating rovers on the surface of the planet to collect samples for retrieval. During that phase of the mission the spacecraft might take up an orbit with a period that kept it stationary above the Martian equator. Fortunately, the orbits of the moons are close

**ABOVE An engineering study of a subsidiary vehicle to enable members of a crew in orbit around Mars to investigate the surfaces of the Martian moons.** *(NASA Langley Research Center and AMA Studios)*

**RIGHT Artwork by Pat Rawlings of humans operating in the vicinity of one of the Martian moons.** *(NASA)*

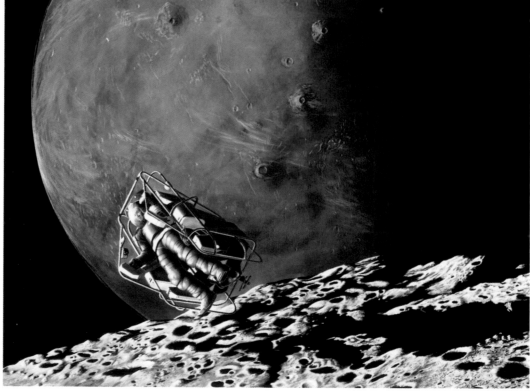

to the equator of the planet. Because Phobos circles Mars in less time than the planet takes to rotate on its axis, and Deimos takes longer, the altitude of a stationary orbit would lie between those of the moons. Hence the mission might start with sampling Phobos, then operations on the planet, and after those samples had been retrieved the focus would switch to Deimos. As a side benefit, that would complete the orbital mission with the vehicle in a good position for the escape manoeuvre to head back to Earth.

# Advanced propulsion

The prospects for human settlement of Mars depend on whether we introduce advanced methods of interplanetary propulsion.

Although we could start the expeditionary phase with chemical rockets, the issue is whether it would be wiser to await the introduction of a more advanced means of interplanetary propulsion. Certainly such a system will be required for long-term settlement of the planet.

There are a number of options.

Solar-electric propulsion draws energy from sunlight to power engines that deliver thrust by accelerating ions drawn from a supply of fuel. This technology has been used by deep space probes, most notably the Dawn mission that was able to visit two asteroids in a manner that would not have been possible for a spacecraft of that size using chemical propulsion. For a human mission to Mars the thrust would have to be much greater, in turn requiring considerably bigger solar panels. Nuclear-electric propulsion is similar but the energy to run the ion drive is obtained from a nuclear fission reactor. An electric drive provides a low thrust and therefore must be maintained for a long time, perhaps a substantial portion of the travel time. One disadvantage is that if such a drive performs the manoeuvre to escape from Earth orbit into interplanetary space, it will spend a long time spiralling out through the region occupied by the van Allen radiation belts. It might be parked in a Lagrange point of the Earth–Moon system and the crew would fly a faster trajectory from low Earth orbit to begin their mission. An ion drive will also have to spiral into orbit around Mars, transfer between orbits, and then escape to head home.

## SATELLITES OF MARS

|  | Phobos | Deimos |
|---|---|---|
| Semi-major axis (km)[1] | 9,378 | 23,459 |
| Sidereal orbit period (days) | 0.31891 | 1.26244 |
| Sidereal rotation period (days)[2] | 0.31891 | 1.26244 |
| Orbital inclination (°) | 1.08 | 1.79 |
| Orbital eccentricity | 0.0151 | 0.0005 |
| Radius on axis facing Mars (km) | 13.0 | 7.8 |
| Radius in direction of travel (km) | 11.4 | 6.0 |
| Axis in polar direction (km) | 9.1 | 5.1 |
| Mass ($10^{15}$ kg) | 10.6 | 2.4 |
| Mean density (kg/m$^3$) | 1,900 | 1,750 |
| Geometric albedo | 0.07 | 0.08 |

(Data courtesy of NASA)
Notes: (1) The mean orbital distance from the centre of Mars. (2) The moons are tidally locked, so their axial rotations are synchronised with their orbits.

In the case of nuclear-thermal propulsion, the fission heats a fluid such as liquid hydrogen, causing it to expand and exit from a nozzle as thrust. In effect, this is similar to a chemical rocket except that the chemical energy liberated by the interaction of the propellants is replaced by the heat of the nuclear reaction. Such an engine can provide a high thrust to attain a desired trajectory, then the spacecraft will coast. When not thrusting, the spacecraft can power its systems either by running the reactor at low level or by deploying solar panels. The high thrust that such an engine can deliver will eliminate the long periods that an ion drive will spend in a spiralling path. Furthermore, if the high thrust is maintained for longer, it will trim the time to reach Mars by an order of magnitude, perhaps even reducing the one-way trip to several weeks.

Anything that involves nuclear energy tends to attract criticism, but if ever there was an environment where it was safe to operate a fission reactor, that's deep space because it is already seething with radiations of all kinds. The task will simply be to shield the crew and there are various ideas for achieving that.

If we manage to achieve self-sustaining nuclear fusion and miniaturise the technology so that it can drive a rocket through interplanetary space, we might make science fiction a reality with people hopping between planets in the same casual manner as we now fly between continents.

# Postscript

**E**arly on, we were fascinated by Mars for its blood-red hue. Then we studied its motions in the sky, from which we learned how planets orbit the Sun.

With the invention of the telescope, Mars became a world with features on its face that allowed us to determine its period of rotation and its seasonal variations. As telescopes improved we drew ever more detailed maps. Some observers saw networks of straight lines, and speculated that they were made by a dying race of intelligent beings. This idea was eventually discarded but the general feeling was that Mars hosted some form of vegetation.

Then flyby space probes revealed the planet to be heavily cratered, with an ancient, inert surface. This impression was soon overturned by orbital mapping, which showed the planet to have been geologically active. It had mountains far taller than Mount Everest, with craters on their summits establishing them to be ancient volcanoes. There was a canyon system considerably longer and deeper than the Grand Canyon. And there were broad channels cut by floods far larger than any on Earth. All of this was fascinating for geologists, but what about life?

The Viking landers, equipped to answer the question of whether there was microbial life in the soil, definitively proved neither its presence nor its absence.

Later, orbiters carrying multispectral cameras and numerous other sensors mapped minerals at the surface and water in the near-subsurface. Gradually we became aware of how conditions changed. Early on, the atmosphere was much thicker and conditions were warmer and wetter.

Now rovers are trundling across the surface, acting as field geologists and seeking evidence of environments that would once have been conducive to life, particularly ancient lakes.

Highly capable though they are, robots are just precursors to direct human participation in the exploration of the planet.

A human landing on Mars is a staple of science fiction that may well soon become reality. We cannot yet say when the first boot print will be made in the Martian soil but it seems likely this person is alive today. Perhaps this book will help to stimulate their pursuit of that career path!

**OPPOSITE Sunset on Mars imaged by the Spirit rover. One day humans will be able to witness this for themselves.** *(NASA/JPL-Caltech/Texas A&M Univ./Cornell)*

# Maps of Mars

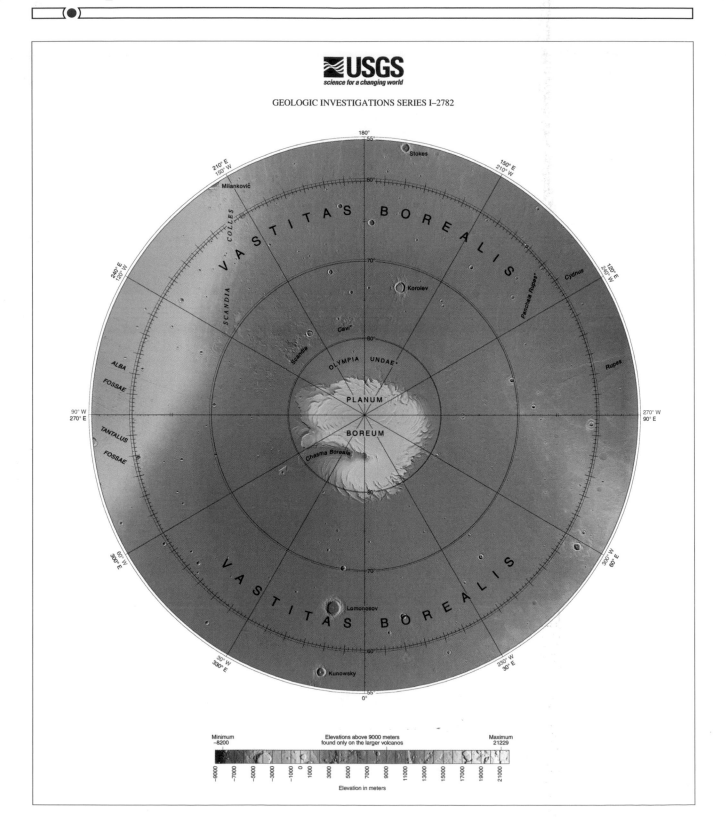

**≋USGS**
science for a changing world

GEOLOGIC INVESTIGATIONS SERIES I–2782

Minimum
−8200

Elevations above 9000 meters
found only on the larger volcanos

Maximum
21229

Elevation in meters

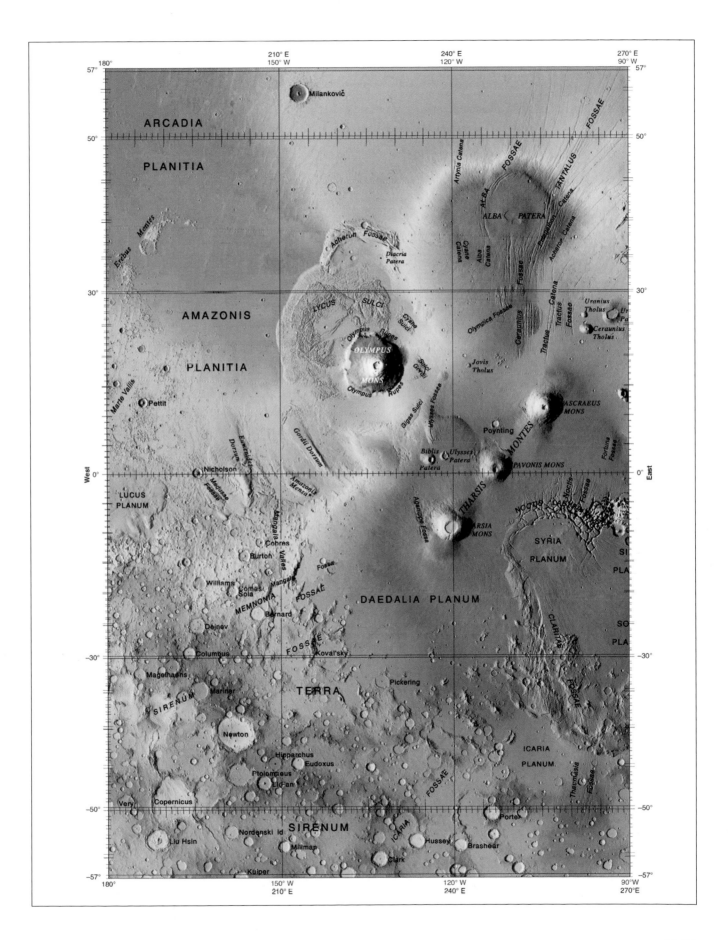

ARCADIA

PLANITIA

Milanković

Erebus Montes

AMAZONIS

PLANITIA

Marte Vallis

Pettit

Nicholson

LUCUS
PLANUM

Medusae Fossae

Mangala Valles

Cobres

Burton

Williams

Comas Sola

Dejnev

Columbus

MEMNONIA

FOSSAE

Magelhaens

SIRENUM

Mariner

Newton

Hipparchus

Ptolomaeus

Li Fan

Very

Copernicus

Liu Hsin

Nordenski Id

Kuiper

Millman

SIRENUM

Eudoxus

TERRA

Pickering

Koval'sky

Bernard

Fossa

Mangala

Gordii Dorsum

Enenchilus Dorsum

Amazonis Mensa

Acheron Fossae

Diacria Patera

LYCUS SULCI

Olympus Rupes

Cyane Sulci

OLYMPUS MONS

Olympus Rupes

Sulci Gordii

Gigas Sulci

Ulysses Fossae

Biblis Patera

Ulysses Patera

Agenippe Fossae

DAEDALIA PLANUM

ARSIA MONS

ICARIA

FOSSAE

Clark

Hussey

Brashear

Porter

ICARIA

PLANUM

Thaumasia Fossae

Artynia Catena

ALBA FOSSAE

TANTALUS FOSSAE

ALBA PATERA

Cyane Catena

Alba Catena

Ceraunius Fossae

Olympica Fossae

Tractus Catena

Tractus Fossae

Jovis Tholus

Poynting

Uranius Tholus

Ur Pa

Ceraunius Tholus

ASCRAEUS MONS

Fortuna Fossae

PAVONIS MONS

THARSIS MONTES

SYRIA PLANUM

NOCTIS LABYRINTHUS

Noctis Fossae

CLARITAS FOSSAE

SI PLA

SO PLA

West

East

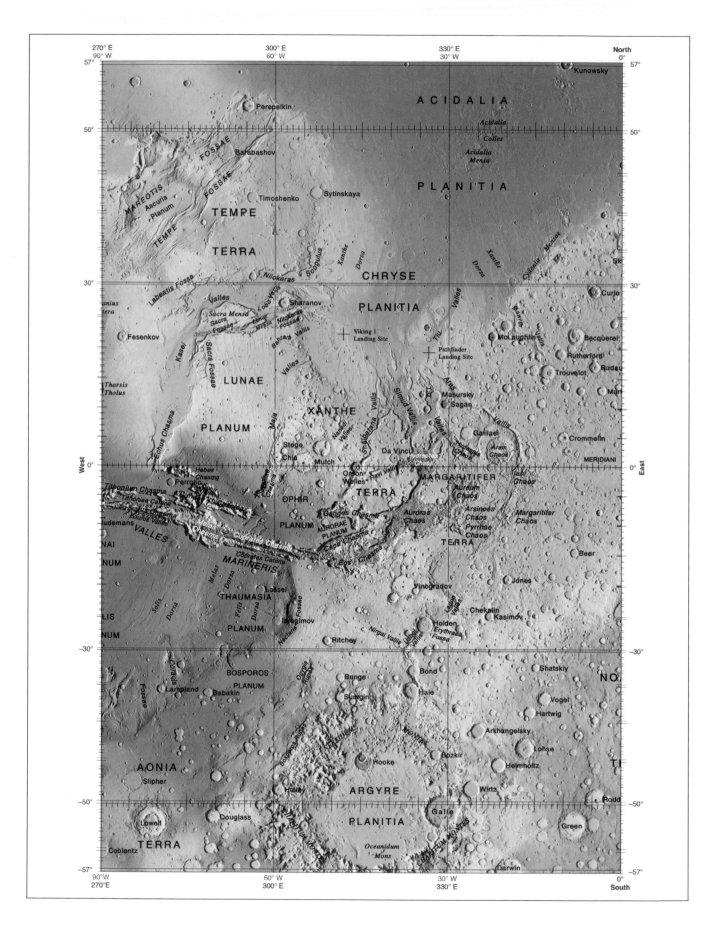

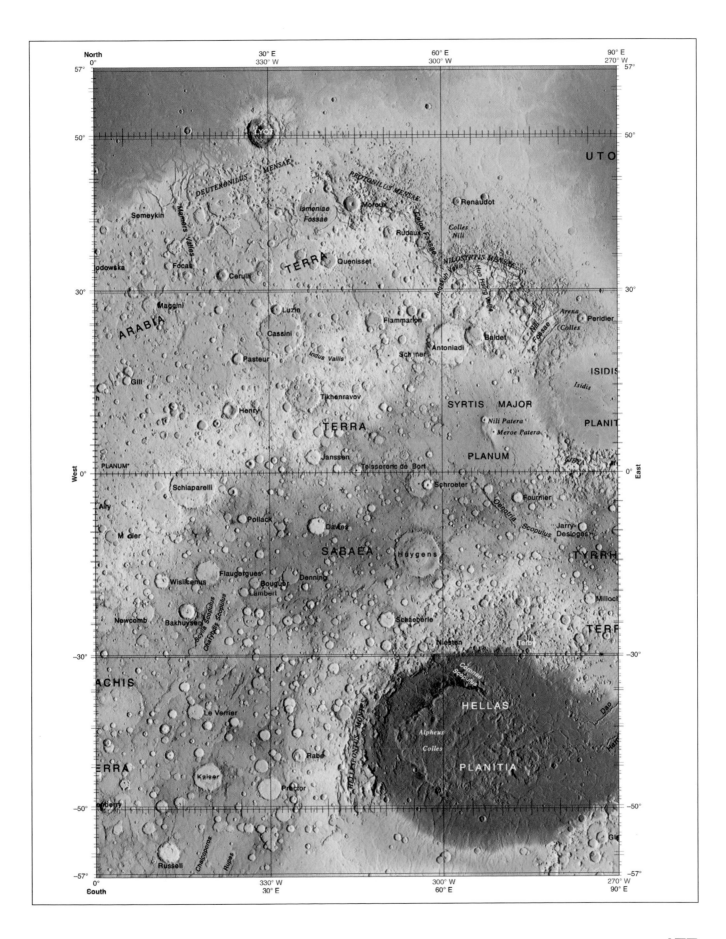

North
0°

30° E
330° W

60° E
300° W

90° E
270° W

57°                                                                                              57°

50°                                    LYOT                                                    50°

UTO

*DEUTERONILUS MENSAE*

*PROTONILUS MENSAE*

Renaudot

Semeykin          *Ismeniae*          Moreux
                  *Fossae*
                                                    *Colles*
*Mamers Vallis*                    Rudaux            *Nili*

                                                    *NILOSYRTIS MENSAE*

Iodowska          Focas                                              *Auxakuh Vallis*   *Huo Hsing Vallis*

              Cerulli          T E R R A    Quenisset                                    *Nili*

30°                                                                                              30°

Maggini                                                              *Arena*   Peridier
                          Luzin                      Flammarion                 *Colles*
ARABIA
                  Cassini                                            Baldet
                                                Schiner   Antoniadi        *Nili*
        Pasteur                  *Indus Vallis*                                *Fossae*
                                                                                        ISIDIS
   Gill                                                              *Isidis*
 h
                          Tikhonravov          T E R R A                    SYRTIS   MAJOR     PLANIT

                                                            *Nili Patera*                    PLANIT
                  Henry                                      *Meroe Patera*

              T E R R A                Janssen                                    PLANUM

West  PLANUM*                              Teisserenc de Bort                          East
0°                                                                                              0°

              Schiaparelli                      Schroeter
                                                                    Fournier
   Airy
                      Pollack              Davies                              Jarry-
 M dier                                                                        Desloges
                                                            *Oenotria Scopulus*
                                  S A B A E A      Huygens                          TYRRH
        Wislicenus      Flaugergues
Newcomb          Bouguer      Denning                                              Millocl
                  Lambert                                                              TERR
        Bakhuysen                                    Schaeberle
                  *Boríla Scopulus*
                  *Oharipdis Scopulus*                  Niesten          Terby
-30°                                                                                              -30°
ACHIS
                                                            *Coronae Scopulus*
                                              HELLAS                    *Dao*
                  Le Verrier
 ERRA                                          *Alpheus*              *Hadr*
                                                    *Colles*
                          Rabe
                                              PLANITIA
        Kaiser
                                                              *Hellespontus Montes*

 ferry          Proctor                                                          Gl
-50°                                                                                              -50°

        Russell   *Chalcoporos Rupes*

-57°                                                                                              -57°
0°
South

330° W
30° E

300° W
60° E

270° W
90° E

**177**
MAPS OF MARS

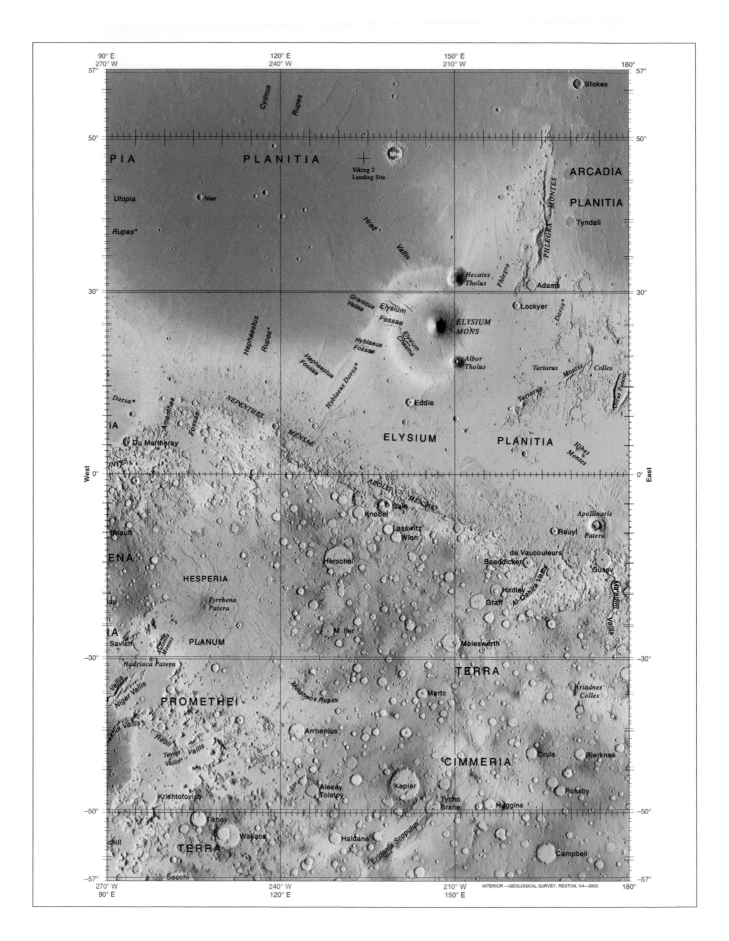

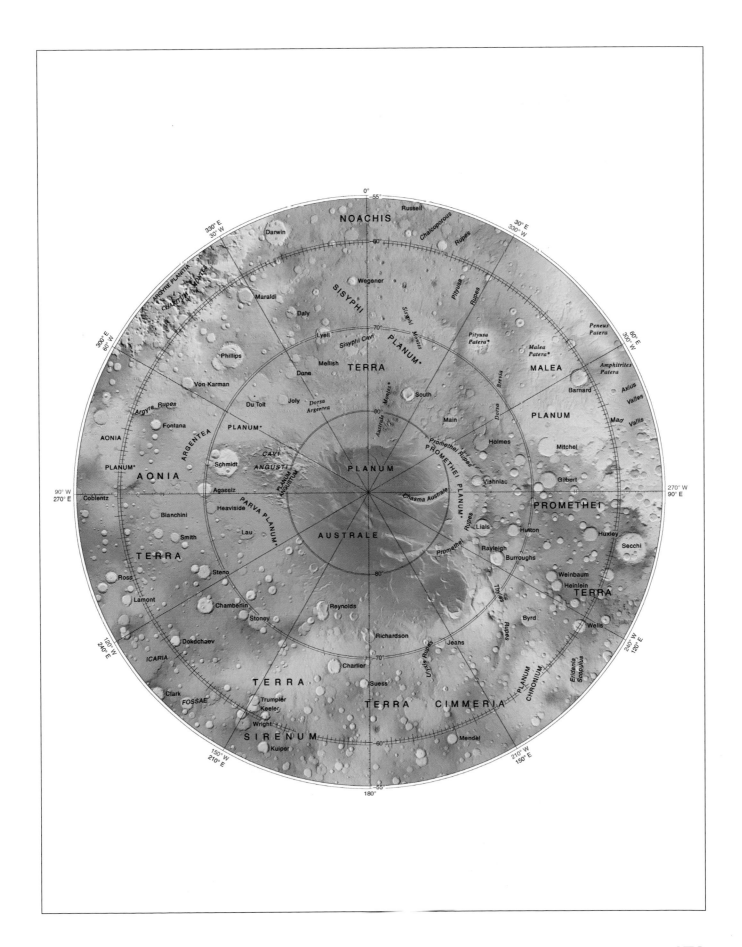

# Mission chronology

| Name | Nation | Launch | Type | Outcome |
|---|---|---|---|---|
| – | USSR | 10 October 1960 | Flyby | Failed to achieve Earth parking orbit. |
| – | USSR | 14 October 1960 | Flyby | Failed to achieve Earth parking orbit. |
| – | USSR | 24 October 1962 | Flyby | Stranded in Earth parking orbit. |
| Mars 1 | USSR | 1 November 1962 | Flyby | Spacecraft suffered attitude control difficulties that prevented a midcourse correction. It fell silent on 21 March 1963, 106 million km from Earth. It flew past Mars on 19 June 1963 at a range of almost 200,000km. |
| – | USSR | 4 November 1962 | Atmosphere probe | Stranded in Earth parking orbit. The mission was to have delivered a probe to report conditions in the Martian atmosphere, to help to design future landers. |
| Mariner 3 | USA | 5 November 1964 | Flyby | Payload shroud failed to separate. |
| Mariner 4 | USA | 28 November 1964 | Flyby | Flew past Mars on 15 July 1965 at a range of just under 10,000km and returned 22 images (the later ones in darkness). |
| Zond 2 | USSR | 30 November 1964 | Flyby | It fell silent in transit and then flew past Mars on 6 August 1965. The encounter range is disputed. At issue is whether control of the spacecraft was lost very early in 1965 or shortly after a second course correction it would have undertaken in May. The Soviets reported on 5 May that contact had been lost. Contemporary Russian sources specified the range as 650,000km but some Westerners said it might have been as close as 1,500km. The plan may have been to make such a close flyby, but to achieve it the craft would have had to carry out a manoeuvre in May. |
| Zond 3 | USSR | 18 July 1965 | Development test flight | It tested the imaging system during a flyby of the Moon on 20 July 1965. Transmitted the pictures at different interplanetary distances to assess quality. Fell silent on 3 March 1966 when at a distance of 150 million km. |
| Mariner 6 | USA | 25 February 1969 | Flyby | Flew past Mars on 31 July 1969 at a range of 3,431km and returned images. |
| – | USSR | 27 March 1969 | Orbiter | Failed to achieve Earth parking orbit. |
| Mariner 7 | USA | 27 March 1969 | Flyby | Flew past Mars on 5 August 1969 at a range of 3,430km and returned images. |
| – | USSR | 2 April 1969 | Orbiter | Failed to achieve Earth parking orbit. |
| – | USSR | 10 May 1971 | Orbiter | Stranded in Earth parking orbit. It was meant to have flown a fast trajectory to Mars and entered orbit ahead of its two companions, then act as a radio beacon to optimise their approach prior to releasing their landers. |
| Mars 2 | USSR | 19 May 1971 | Orbiter | In the absence of the intended radio beacon, on 21 November 1971 it used optical navigation and refined its course. On 27 November it performed another adjustment and released its lander, then moved off the impact trajectory ready for its orbit insertion burn 4.5hr later. The period of the orbit was 18hr instead of 24hr. The inclination was 49°. It suffered communications problems, and in any case imaging was frustrated by the dust storm. It exhausted its attitude control system in July 1972. |
| Mars 2 | USSR | 19 May 1971 | Direct entry lander | It entered the atmosphere on 27 November 1971 at a steeper angle than planned and it struck the surface before it could deploy its parachute. The impact was calculated to have occurred at about 44°S, 313°W. |
| Mars 3 | USSR | 28 May 1971 | Orbiter | On 2 December 1971 it used optical navigation to adjust its trajectory and released its lander. When performing its orbit insertion, the burn was shorter than planned and the period of the highly elliptical 49° orbit was 12.8 days, offering few opportunities for imaging. In any case this was frustrated by the dust storm. It exhausted its attitude control system in July 1972. |

| Name | Nation | Launch | Type | Outcome |
|------|--------|--------|------|---------|
| Mars 3 | USSR | 28 May 1971 | Direct entry lander | Landed successfully on 2 December 1971, but fell silent 14.5sec later when preparing an image. The site was at 45°S, 158°W. |
| Mariner 8 | USA | 9 May 1971 | Orbiter | Failed to achieve Earth parking orbit. |
| Mariner 9 | USA | 30 May 1971 | Orbiter | Entered orbit on 14 November 1971 inclined at 64° to the equator. Mapping was postponed until after the dust storm abated. Deactivated after 516 days having consumed its attitude control propellant. It returned some 7,000 pictures and mapped 70% of the planet. |
| Mars 4 | USSR | 21 July 1973 | Orbiter | Failed to execute the orbit insertion burn and flew past the planet on 10 February 1974 at a range of 1,800km. |
| Mars 5 | USSR | 25 July 1973 | Orbiter | On 12 February 1974 it achieved an orbit with a period of 24.88hr inclined at 35°. The instrument compartment developed a leak that meant loss of communication after only 22 orbits. |
| Mars 6 | USSR | 5 August 1973 | Flyby/direct entry lander | On 12 March 1974 the lander hit the surface at 24°S, 19.5°W at about 60m/sec, too fast for its survival. It transmitted data during the descent, which was relayed by the bus while making its 1,600km flyby. |
| Mars 7 | USSR | 9 August 1973 | Flyby/direct entry lander | Although the bus released the lander on 9 March 1974 as planned, a malfunction caused it to miss the planet by 1,300km. The intended landing site was at 51°S, 31°W. |
| Viking 1 | USA | 20 August 1975 | Orbiter | On 19 June 1976 it entered an orbit inclined 40° to the equator, and on 21 June it tweaked the period to 24.66hr to match that of the planet's rotation. It released the lander on 20 July. It was deactivated on 7 August 1980, having exhausted its supply of attitude control propellant. |
| Viking 1 | | | Lander | Touched down on 20 July 1976 on Chryse Planitia at 22.697°N, 48.222°W. It was named the Thomas Mutch Memorial Station in January 1982. Contact was maintained until a faulty command uplink on 11 November 1982 resulted in loss of contact. |
| Viking 2 | USA | 9 September 1975 | Orbiter | On 7 August 1976 it entered an orbit inclined 55° and on 9 August tweaked the period to 27.3hr. It released the lander on 3 September. It raised its inclination to 75° on 30 September and to 80° on 20 December. It suffered a leak of attitude control propellant and was switched off on 25 July 1978. |
| Viking 2 | | | Lander | It touched down on 3 September 1976 on Utopia Planitia at 48.269°N, 225.990°W. It was named the Gerald A. Soffen Memorial Station. With its battery failing, it was deactivated on 12 April 1980. |
| Fobos 1 | USSR | 7 July 1988 | Orbiter | It fell silent in transit after an erroneous command uplink on 29 August 1988. This was realised on 2 September when the spacecraft failed to respond to a scheduled radio call. |
| Fobos 2 | USSR | 12 July 1988 | Orbiter | On 29 January 1989 it entered an equatorial orbit with a period of 78hr. By 18 February it was in an orbit slightly higher than Phobos, with a period a few minutes longer. After several flybys and some orbital tweaks it fell silent on 27 March during the final preparations to drop instrument payloads on the moon. The problem was attributed to a flaw in the computer system. |
| Mars Observer | USA | 25 September 1992 | Orbiter | Fell silent on 21 August 1993 while preparing its propulsion system for the orbit insertion burn. |
| Mars Global Surveyor | USA | 7 November 1996 | Orbiter | Entered orbit on 12 September 1997 inclined at 93° to the equator (Sun-synchronous). After 552 sols of aerobraking it achieved its operating orbit, with a period of 2hr arranged always to cross the day-side equator at 2pm (local time) moving from south to north. Successive orbits were 28.62° to the west due to the revolution of Mars. After 7 sols and 88 orbits, the spacecraft would approximately retrace its previous path, with an offset of 59km to the east, ensuring eventual coverage of the entire surface. It fell silent on 2 November 2006 after an erroneous command uplink. |
| Mars 8 | Russia | 16 November 1996 | Orbiter | Stranded in Earth parking orbit. |
| Mars Pathfinder and Sojourner | USA | 4 December 1996 | Direct entry lander | It landed on 4 July 1997 on Chryse Planitia in the outflow from Ares Vallis at 19.13°N, 33.22°W. The rover, Sojourner, drove onto the surface on sol 2. The lander fell silent on 27 September 1997. This denied the still operating rover its relay with Earth. It was named the Carl Sagan Memorial Station. |

| Name | Nation | Launch | Type | Outcome |
|---|---|---|---|---|
| Nozomi (Hope) | Japan | 3 July 1998 | Orbiter | A propellant leak during the Earth escape burn prevented it attempting Mars orbit insertion, with the result that on 14 December 2003 it flew past the planet at a range of 1,000km. |
| Mars Climate Orbiter | USA | 11 December 1998 | Orbiter | A procedural flaw caused a navigational error that resulted in the spacecraft deeply penetrating the atmosphere of Mars during the insertion process and burning up on 23 September 1999. |
| Mars Polar Lander | USA | 3 January 1999 | Direct entry lander and penetrator probes | Failed to land on 3 December 1999. Nothing was ever heard from the Deep Space 2 probes named Scott and Amundsen that were released to enter the atmosphere independently and penetrate the surface. |
| Mars Odyssey | USA | 7 April 2001 | Orbiter | On 24 October 2001 it entered a Sun-synchronous orbit inclined at 93°. After three months of aerobraking it achieved its operating orbit. It is still active.[1] |
| Mars Express | Europe | 2 June 2003 | Orbiter | It released the Beagle 2 lander on 19 December 2003. The spacecraft entered an orbit inclined at 25° to the equator on 25 December, then made a series of manoeuvres to raise the angle to 86° in preparation for trimming the period to 6.7hr. It is still active. |
| Beagle 2 | Europe | 2 June 2003 | Direct entry lander | On 25 December the Beagle 2 probe entered the atmosphere, but there was no signal. It was later found to have landed successfully but some of its solar panels failed to unfold, preventing it from communicating. It touched down on Isidis Planitia at 11.5265°N, 90.4295°E. |
| MER-A and Spirit | USA | 10 June 2003 | Direct entry lander with a rover | It touched down on 4 January 2004 in Gusev at 14.5684°S, 175.472636°E. The site was named the Columbia Memorial Station. The Spirit rover was deployed on sol 12. The last communication with the rover was on 22 March 2010, after becoming bogged down in soft soil. |
| MER-B and Opportunity | USA | 8 July 2003 | Direct entry lander with a rover | It touched down on 25 January 2004 on Meridiani Planum at 1.9462°S, 354.4734°E. The site was named the Challenger Memorial Station. Having gained confidence in the procedures, it deployed the Opportunity rover on sol 7. It has driven over 50km. It is still active. |
| Rosetta | Europe | 2 March 2004 | Flyby gravity assist | Flew past Mars on 25 February 2007 at a range of only 250km (a risky flyby nicknamed 'The Billion Euro Gamble') in transit to rendezvous with comet 67P/Churyumov-Gerasimenko. |
| Mars Reconnaissance Orbiter | USA | 12 August 2005 | Orbiter | Entered orbit on 10 March 2006 inclined at 93° to the equator (Sun-synchronous). After five months of aerobraking it achieved its operating orbit. It is still active. |
| Mars Phoenix | USA | 4 August 2007 | Direct entry lander | It touched down successfully on 25 May 2008 on Vastitas Borealis at 68.22°N, 125.7°W. Its science mission ended in August. The final communication with Earth was on 2 November as available solar power declined with the onset of northern winter. |
| Dawn | USA | 27 September 2007 | Flyby gravity assist | It flew past Mars on 18 February 2009 at a range of 542km in transit to asteroid 4 Vesta. |
| Fobos-Grunt (Soil) | Russia | 8 November 2011 | Orbiter | Stranded in Earth parking orbit. |
| Yinghuo-1 (Firefly) | China | 8 November 2011 | Orbiter | Lost with Fobos-Grunt. |
| Curiosity | USA | 26 November 2011 | Direct entry skycrane | It deposited the Curiosity rover onto the surface in Gale crater at 4.5895°S, 137.4417°E on 6 August 2012. The site was named Bradbury Landing. The rover is still active. |
| Mangalyaan (Mars Orbiter Mission) | India | 5 November 2013 | Orbiter | Entered a highly eccentric orbit on 24 August 2014 inclined 150° to the equator. It is still active. |
| MAVEN | USA | 18 November 2013 | Orbiter | Entered an orbit inclined at 75° to the equator on 22 September 2014. It is still active. |
| Trace Gas Orbiter | Europe | 14 March 2016 | Orbiter | After releasing Schiaparelli on 16 October 2016 it achieved an orbit inclined at 74° to the equator on 19 October. It is still active. |
| Schiaparelli | Europe | 14 March 2016 | Direct entry lander | Crashed on 19 October 2016 attempting to land on Meridiani Planum at 2.07°S, 6.21°W. |

**ABOVE A true-colour view of Mars taken on 24 February 2007 by the Rosetta spacecraft while flying past the planet.**
*(ESA/MPS for OSIRIS Team)*

| Name | Nation | Launch | Type | Outcome |
|---|---|---|---|---|
| **FORTHCOMING** | | | | |
| InSight | USA | (2018) | Direct entry lander | It is to operate seismometers and drill to emplace heat-flow sensors. |
| Mars 2020 | USA | (2020) | Direct entry skycrane | It is to deposit a rover onto the surface. |
| ExoMars 2020 | Europe | (2020) | Direct entry lander | It is to deploy a rover. |

(Data courtesy of NASA, ESA, ISRO and Wes Huntress)

Note: (1) The comment 'still active' refers to December 2017, when this book was completed.

# Mars compared to Earth

| | Mars | Earth | Mars/Earth |
|---|---|---|---|
| Semi-major axis ($10^6$km) | 227.939 | 149.60 | 1.524 |
| Perihelion ($10^6$km) | 206.62 | 147.09 | 1.405 |
| Aphelion ($10^6$km) | 249.23 | 152.10 | 1.639 |
| Mean orbital velocity (km/sec) | 24.07 | 29.78 | 0.808 |
| Max. orbital velocity (km/sec) | 26.50 | 30.29 | 0.875 |
| Min. orbital velocity (km/sec) | 21.97 | 29.29 | 0.750 |
| Orbit inclination to ecliptic (°) | 1.850 | 0 | – |
| Orbit eccentricity | 0.0935 | 0.0167 | 5.599 |
| Sidereal rotation period (hr) | 24.6229 | 23.9345 | 1.029 |
| Length of local day (hr) | 24.6597 | 24.0 | 1.027 |
| Sidereal orbit period (days) | 686.980 | 365.256 | 1.881 |
| Local days/year | 668.599 | 365.256 | 1.830 |
| Obliquity of equator to orbit (°) | 25.19 | 23.439 | 1.075 |
| Mass ($10^{24}$kg) | 0.64171 | 5.9724 | 0.107 |
| Volume ($10^{10}$km$^3$) | 16.318 | 108.321 | 0.151 |
| Equatorial radius (km) | 3,396.2 | 6,378.1 | 0.532 |
| Polar radius (km) | 3,376.2 | 6,356.8 | 0.531 |
| Volumetric mean radius (km) | 3,389.5 | 6,371.0 | 0.532 |
| Core radius (km) | 1,700 | 3,485 | 0.488 |
| Flattening | 0.00589 | 0.00335 | 1.76 |
| Surface area ($10^6$km$^2$)[1] | 144.798 | 510.072 | 0.284 |
| Mean density (kg/m$^3$) | 3,933 | 5,514 | 0.713 |
| Surface gravity (m/sec$^2$) | 3.711 | 9.807 | 0.378 |
| Escape velocity (km/sec) | 5.027 | 11.186 | 0.450 |
| Solar irradiance (W/m$^2$) | 586.2 | 1,361 | 0.431 |
| Topographic range (km) | 30 | 20 | 1.5 |
| Surface pressure (kPa)[2] | 0.636 | 101.325 | 0.006 |

(Data courtesy of NASA)

Notes: (1) Of Earth's surface area of 510 million sq km, 149 million sq km is land and the rest is water, therefore the surface area of Mars is almost the same as that of Earth. (2) See ftable on page 186 for further details.

**RIGHT** Size comparison of **Earth and Mars.** *(NASA)*

**The 70m dish of the Deep Space Network at Goldstone in California is known as the Mars antenna.**
*(NASA/JPL-Caltech)*

# The atmosphere of Mars

| Surface pressure (kPa)[1] | – | 0.636 |
|---|---|---|
| Scale height (km) | – | 11.1 |
| Total mass of atmosphere ($10^6$kg) | – | approx. 2.5 |
| Mean molecular weight | – | 43.34 |
| Composition by volume | carbon dioxide | 95.32% |
| – | molecular nitrogen | 2.7% |
| – | argon | 1.6% |
| – | molecular oxygen | 0.13% |
| – | carbon monoxide | 0.08% |
| – | water | 210ppm |
| – | nitrogen oxide | 100ppm |
| – | neon | 2.5ppm |
| – | hydrogen-deuterium oxide | 0.85ppm |
| – | krypton | 0.3ppm |
| – | xenon | 0.08ppm |

(Data from the Viking landers, courtesy of NASA)

Note: (1) The mean pressure at the surface of Mars is 0.636kPa, but ranges from 0.030kPa atop Olympus Mons to 1.155kPa on the floor of the Hellas basin.

**BELOW The range of temperatures on Earth and Mars.** (NASA)

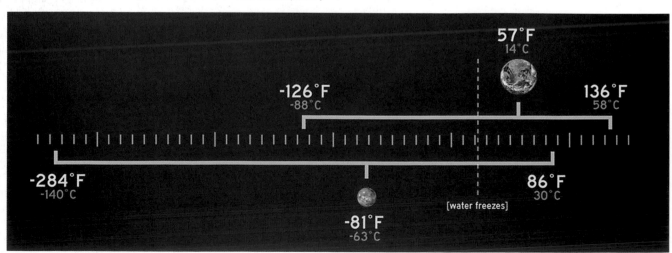

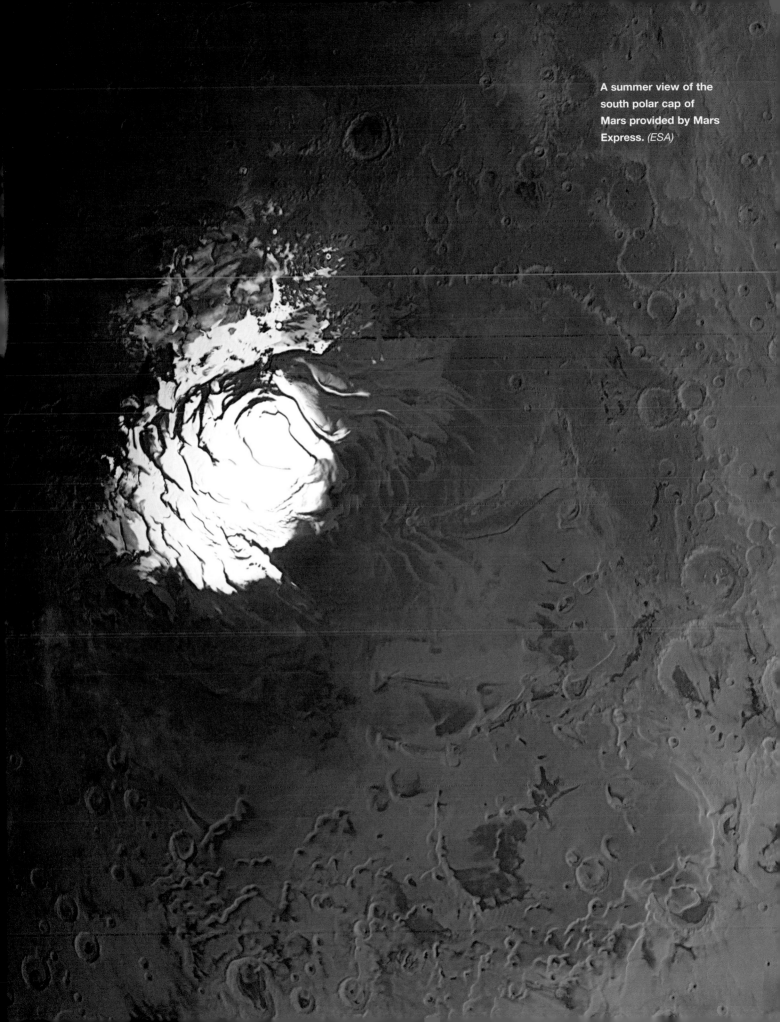

A summer view of the south polar cap of Mars provided by Mars Express. *(ESA)*

# Further reading

*(in chronological order)*

Percival Lowell, *Mars*, Longmans, Green & Co., 1895

Herbert George Wells, *The War of the Worlds*, Heinemann, 1898

Percival Lowell, *Mars and its Canals*, Macmillan, 1906

Alfred Russel Wallace, *Is Mars Habitable?*, Macmillan, 1907

Percival Lowell, *Mars as the Abode of Life*, Macmillan, 1908

Edgar Rice Burroughs, *A Princess of Mars*, A. C. McClurg, 1917

Robert A. Heinlein, *Red Planet*, Scribner's, 1949

Ray Bradbury, *The Martian Chronicles*, Doubleday, 1950

Gerard de Vaucouleurs, *The Planet Mars*, translated by Patrick Moore, Faber and Faber, 1951

Arthur C. Clarke, *The Sands of Mars*, Sidgwick and Jackson, 1951

Wernher von Braun, *The Mars Project*, University of Illinois, 1953

Patrick Moore, *Guide to Mars*, Frederick Muller, 1956 (revised edition 1965)

R. S. Richardson, *Man and the Planets*, Scientific Book Club, 1956

Wernher von Braun and Willy Ley, *The Exploration of Mars*, Viking, 1956

Howard Koch, *The Panic Broadcast*, Avon, 1970

William K. Hartmann and Odell Raper, *The New Mars: The Discoveries of Mariner 9*, NASA-SP-337, 1974

NASA, *Mars as Viewed by Mariner 9*, NASA-SP-329, 1974

Eugene Michael Antoniadi, *The Planet Mars*, translated by Patrick Moore, Reid, 1975

William Graves Hoyt, *Lowell and Mars*, University of Arizona, 1976

Mark Washburn, *Mars At Last! The Red Planet Revealed, from Man's First Sighting to the Viking Touchdown*, Abacus, 1977

NASA, *On Mars Exploration of the Red Planet 1958-1978*, NASA-SP-4212, 1984

Norman H. Horowitz, *To Utopia and Back: The Search for Life in the Solar System*, W. H. Freeman, 1986

Kim Stanley Robinson, *Red Mars*, HarperCollins, 1992

Kim Stanley Robinson, *Green Mars*, HarperCollins, 1993

Kim Stanley Robinson, *Blue Mars*, HarperCollins, 1996

Stephen Baxter, *Voyage*, HarperCollins, 1996

William Sheehan, *The Planet Mars: A History of Observation and Discovery*, University of Arizona, 1996

Robert Zubrin and Richard Wagner, *The Case for Mars: The Plan to Settle the Red Planet and Why We Must*, Free Press, 1997

Martin Caidin, Jay Barbree and Susan Wright, *Destination Mars: In Art, Myth and Science*, Penguin, 1997

Malcolm Walter, *The Search for Life on Mars*, Allen and Unwin, 1999

Paul Chambers, *Life On Mars: The Complete Story*, Blandford, 1999

Laurence Bergreen, *The Quest for Mars: The NASA Scientists and their Search for Life Beyond Earth*, HarperCollins, 2000

David S. Portree, *Humans to Mars: Fifty Years of Mission Planning 1950-2000*, NASA-SP-4521, 2001

William Sheehan and Stephen James O'Meara, *Mars: The Lure of the Red Planet*, Prometheus, 2001

Oliver Morton, *Mapping Mars: Science, Imagination and the Birth of a World*, Fourth Estate, 2002

Joseph M. Boyce, *The Smithsonian Book of Mars*, Smithsonian Institution Press, 2002

William K. Hartmann, *A Traveler's Guide to Mars: The Mysterious Landscapes of the Red Planet*, Workman, 2003

Andrew Mishkin, *Sojourner: An Insider's View of the Mars Pathfinder Mission*, Berkley, 2003

Colin Pillinger, *Beagle: From Darwin's Epic Voyage to the British Mission to Mars*, Faber and Faber, 2003

Michael Hanlon, *The Real Mars: Spirit, Opportunity, Mars Express and the Quest to Explore the Red Planet*, Constable, 2004

Tetsuya Tokano, *Water on Mars and Life*, Springer, 2005

Steve Squyres, *Roving Mars: Spirit, Opportunity, and the Exploration of the Red Planet*, Hyperion, 2005

Andrew Chaikin, *A Passion for Mars: Intrepid Explorers of the Red Planet*, Abrams, 2008

Donald Rapp, *Human Missions to Mars: Enabling Technologies for Exploring the Red Planet*, Springer, 2008

Wesley T. Huntress and Mikhail Ya. Marov, *Soviet Robots in the Solar System: Mission Technologies and Discoveries*, Springer, 2011

Philip J. Stooke, *The International Atlas of Mars Exploration: The First Five Decades*, Cambridge University, 2012

David Baker, *NASA Mars Rovers Manual: 1997-2013 (Sojourner, Spirit, Opportunity and Curiosity)*, Haynes, 2013

Camille Flammarion, *The Planet Mars*, French 1892 edition translated by Patrick Moore and edited by William Sheehan, Springer, 2015

Giancarlo Genta, *Next Stop Mars: The Why, How, and When of Human Missions*, Springer, 2017

Manfred "Dutch" von Ehrenfried, *Exploring the Martian Moons: A Human Mission to Deimos and Phobos*, Springer, 2017

-o0o-

*Readers seeking further coverage of given missions should consult the Planetary Image Archive hosted by NASA at http://photojournal.jpl.nasa.gov.*